PLEASE RETURN THIS ITEM
BY THE DUE DATE TO ANY
TULSA CITY-COUNTY LIBRARY.

FINES ARE 5¢ PER DAY; A
MAXIMUM OF $1.00 PER ITEM.

SUPERNOVA!

CHRISTOPHER LAMPTON

Franklin Watts/An Impact Book
New York/London/Toronto/Sydney/1988

Illustrations by Vantage Art

Photographs courtesy of: Photo Researchers, Inc.: pp. 2 (David A. Haroy/SPL), 10 (Fred Espenak/SPL), 43 (bottom, NRAO/AUI/SPL), 69 (Wards Sci/Science Source), 91 (SPL), 99 (Ronald Royer/SPL), 108 (IMB Collaboration/SPL); Carnegie Institution of Washington: pp. 11, 12; National Optical Astronomy Observatories: pp. 13, 101; University of Toronto: p. 15; NASA: pp. 18 (top & center), 50, 79 (bottom), 82; Jet Propulsion Laboratory: p. 18 (bottom); The Bettmann Archive: pp. 26, 29, 38; AIP/Niels Bohr Library: pp. 30, 52, 62, 88; American Museum of Natural History: p. 32; Mount Wilson and Palomar Observatories: p. 33; California Institute of Technology: pp. 34, 39 (William C. Miller/Mt. Palomar Observatory); Yerkes Observatory: pp. 41, 42, 44, 55, 66, 96; Burndy Library: p. 43 (top); Trustees of the Science Museum: p. 46; United Nations/Photo by Sygma: p. 60; Lick Observatory: p. 71; Smithsonian Institute: p. 79 (top).

Library of Congress Cataloging-in-Publication Data

Lampton, Christopher.
Supernova! / by Christopher Lampton.
   p.   cm. — (An Impact book)
Bibliography: p.
Includes index.
Summary: Examines the birth and death of the extremely bright exploding stars known as supernovae and their effects on history and science.
ISBN 0-531-10602-0
1. Supernovas—Juvenile literature. [1. Supernovas.] I. Title.
QB843.S95L36 1988
523.8'446—dc19   88-14301 CIP AC

Copyright © 1988 by Christopher Lampton
All rights reserved
Printed in the United States of America
5 4 3 2 1

CONTENTS

Prologue
9

Chapter One
Twinkle, Twinkle, Brand-new Star
21

Chapter Two
How I Wonder What You Are
49

Chapter Three
Like a Diamond in the Sky
65

Chapter Four
The Many Types of Novas
81

Chapter Five
Hot Times in the Magellanic Cloud
95

Chapter Six
Bright Lights, Dark Matter
107

Glossary
115

Sources Used
121

Recommended Reading
124

Index
125

SUPERNOVA!

PROLOGUE

High in the Andes mountains, above the arid deserts of central Chile, the night skies are so clear that the stars seem to twinkle in the inky heavens like jewels. The Milky Way, a mere wisp of interstellar haze when viewed from North America or Europe, is a clear white band from horizon to horizon, and meteors twinkle in the night like fireflies.

Astronomers are drawn to Chile by these skies and by the soaring mountains, where they can build telescopes that rise above the dense ocean of air that blurs the starlight at lower altitudes. The domes of their observatories rise from the mountain ridges like a field of mushrooms.

On the evening of February 23, 1987, an astronomer named Oscar Duhalde stepped out of the dome that held the 40-inch (100-cm) telescope operated by the Las Campanas Observatory and strolled to the nearby cafeteria. He was looking forward to a hot cup of coffee.

the Southern night sky

Las Campanas Observatory in Chile. The small houselike structure between the observatory domes is where Shelton discovered the supernova.

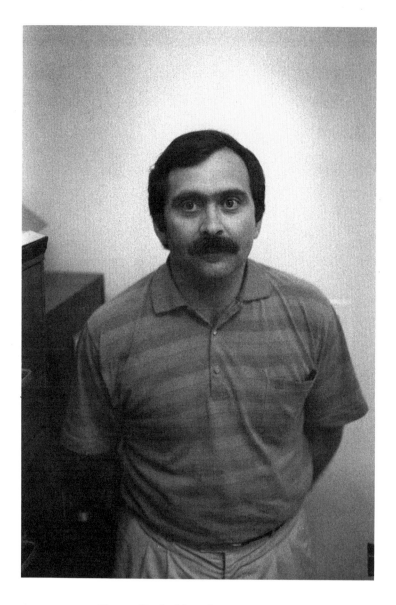

*Oscar Duhalde, the astronomer
who noticed the supernova shortly before
Shelton officially discovered it*

*the Tarantula Nebula, also known
to stargazers as 30 Doradus*

Like most of those working at the observatory, he kept late hours and was absorbed in thought as he allowed his eyes to scan the brilliantly illuminated sky above him.

For a moment his eyes paused at a wisp of light called the Tarantula Nebula, also known as 30 Doradus. He had seen the nebula a hundred times before—a thousand times—and yet now it looked oddly wrong. It seemed brighter than usual. Had something changed? He would have to mention it to someone when he returned to the dome.

But by the time he returned with his cup of coffee, he had forgotten all about the strange light in the sky. . . .

Not far from the dome where Duhalde worked, a young astronomer named Ian Shelton was using a smaller telescope in a smaller dome to take pictures of the sky. By coincidence, the area of the sky that Shelton was photographing was the same one that Duhalde had been observing on his way to get coffee. For three long hours Shelton watched patiently as his photographic plate absorbed the dim light from the stars; when necessary, he turned the telescope by hand to make sure it accurately tracked the motion of the stars through the heavens.

A sharp wind blew across the mountaintop that night. When his exposure was complete, Shelton elected to close the tiny observatory and go to bed. But first he paused in the lab adjacent to the dome to develop the photograph. As Shelton soaked the plate in the baths of pungent chemicals, tiny images of stars appeared against the photographed background of the night sky. Then something odd emerged in the center of the plate, an overexposed smear of light, much brighter than anything he had expected to see. What had gone wrong? For a moment Shelton wondered if there might be a defect in his photographic equipment.

After puzzling over the photograph for long moments, Shelton plunged into the Chilean night and looked at the stars with his own eyes. With amazement, he saw that there was a new star in the sky. It lay within the Large Magellanic Cloud, or LMC, a small, fuzzy patch of light visible only in the skies of the Southern Hemi-

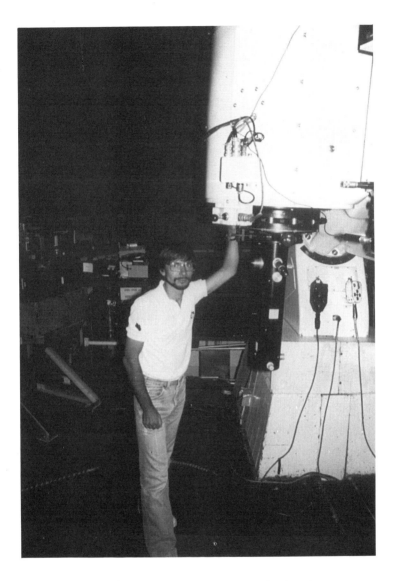

Ian Shelton with his 24-inch (1-m) telescope at Las Campanas. Shelton is employed by the University of Toronto.

sphere. In fact, the LMC is an entire galaxy, a vast cloud of stars not far from the Milky Way Galaxy, of which our own sun is a part.

It was not a terribly bright star; in fact, it was one of the dimmer stars in the sky. Still, Shelton had never seen a star that bright within the LMC. Could he be imagining it? He hurried to the larger dome, where he knew other astronomers would be working. Inside were Duhalde and a second man. He told them what he had seen.

Yes, said Duhalde. He had seen it, too. All at once they all knew what Shelton had seen.

A supernova. Not a new star but an old one that had exploded into sudden brilliance. A star so far away that under normal conditions the human eye could not see it at all. A star in another galaxy flaring as brightly as all the other stars in that galaxy put together.

With a mounting sense of urgency the astronomers tried to telephone the news to the outside world, but the telephone was broken. Duhalde jumped into his car and drove to the nearest town, La Serena, three hours to the north. From there he telegraphed word of the supernova to Cambridge, Massachusetts, in the faraway United States, to the Central Bureau for Astronomical Telegrams.

It was now official. What Shelton had seen was given the name Supernova 1987A. And within hours every astronomer in the Southern Hemisphere had swiveled his or her telescope in the direction of the Large Magellanic Cloud.

In fact, not all of the telescopes turned toward Supernova 1987A were on this planet. The *Solar Max* satellite, placed in orbit around the earth to study the sun, also trained its gamma-ray detector on the supernova, as did the *International Ultraviolet Explorer* satellite, which takes pictures of the sky with a camera sensitive to ultraviolet radiation. And the *Voyager 2* space probe, midway between the planets Uranus and Neptune, was sent instructions to tilt its ultraviolet detectors toward the Large Magellanic Cloud.

What is a supernova? What made the star that Shelton saw in the southern sky—a dim star by normal standards—an occasion for such excitement? Why did astronomers the world over abandon their normal tasks to study this phenomenon? Why did *Time* magazine devote a cover story to it? What is so important—and exciting—about a supernova?

It's a long story. In a sense, the story began long before there were any astronomers on earth to turn their telescopes toward the sky, before there was even an earth for those astronomers to stand on.

Supernovas have played an important role in the history of astronomy and in the history of the universe. Supernovas may have been responsible for the birth of our solar system. They were the crucible in which most of the heavy atoms in your body were forged. Every human being alive has descended, in a very real sense, from one or more supernovas.

According to at least one theory, it was a supernova that inspired the invention of writing. And supernovas were important in helping astronomers throughout history to learn about the universe and our place in it. Supernovas run through the history of astronomy like a golden thread.

And yet, it is one of history's great ironies that the last nearby supernova observed by astronomers occurred *just before* the invention of the telescope, the greatest tool ever developed to study the nature of the heavens.

That is, until Ian Shelton turned his camera on the skies and took a picture of the Large Magellanic Cloud.

Several telescopes in space, including those on the International Ultraviolet Explorer (IUE) *satellite (top), the* Solar Max *satellite (middle), and the* Voyager 2 *spacecraft, were given instructions to take readings and photographs of the supernova.*

1
TWINKLE, TWINKLE, BRAND-NEW STAR

Most of us take the sky for granted. When we look upward on a clear night, we expect to see stars looking down at us. Who but a dedicated astronomer would notice if something changed in the sky from one night to the next?

And yet, truth to tell, the sky is in a constant state of change. Most of the changes are predictable. They occur in cycles. And these cycles are so constant and repetitive that we don't tend to think of them as changes at all.

The most dramatic change, of course, is the change from night to day. During the daytime, the fiery orb that we call the sun passes steadily from horizon to horizon, moving roughly one-twelfth of the way across the sky in the course of an hour. When it sets, the sky turns black and, if there are no clouds, the stars appear. Of course, the stars are also present in the daytime, but we can't see them because of the way the air scatters sunlight to form the familiar blue sky.

During the night, the stars move steadily across the sky, much as the sun moves through the sky during the day, at almost exactly the same speed. After a full night of this motion, the sun rises again from behind the eastern horizon and the cycle begins again.

The sun and stars are not actually moving in a great circle around the earth, of course. What is actually happening is that the earth itself is rotating around an imaginary axis that extends through our planet from the North to the South Pole. This rotation takes twenty-four hours; thus, the sun and the stars seem to swing through the skies on a twenty-four-hour cycle.

But this cycle does not repeat itself exactly every twenty-four hours. Every night the position of the stars changes slightly. This isn't really noticeable on a night-by-night basis, but over a period of weeks it becomes obvious that any given star is rising a little later than it did the previous evening. After six months the stars visible in the sky right after sunset have completely changed.

The changes are caused, of course, by the earth's motion around the sun. Once every 365 days the earth makes one entire orbit around the sun. As the earth moves in its orbit, stars that were previously hidden in the daylight sky emerge into the clear sky of night, and stars that were previously visible at night disappear into the glare of day. Thus, the great sphere of stars makes a complete circuit of the sky every 365 days.

A careful observer, however, will notice that not everything in the sky follows this pattern. The moon, for instance, changes position in the sky every night. In fact, the moon rises an hour later every evening. This is because the moon is orbiting the earth. It takes approximately a month to complete this trip, and thus the moon returns to the same position roughly once a month.

The motion of the earth around the sun also creates a subtler change in the sky that in turn produces weather changes here on earth. As the year progresses, the path that the sun and other objects take through the sky moves both north and south, spending six months moving in one direction and six months moving in the other. Because this changes the angle at which the sun's rays strike the earth, it alters the amount of heat received by various places on the earth's surface. This change in heat in turn produces the changing weather of the four seasons—summer, fall, winter, and spring. The apparent change in the sun's path through the sky is caused by a tilt in the orientation of the earth's axis relative to our orbit around the sun.

There are a handful of starlike objects in the sky that change their position relative to the fixed stars every night. These are the planets. Like the earth, which is itself a planet, they orbit the sun, though most take much longer than an earth year to do so. As they move in their orbits, we see their motion against the fixed backdrop of stars. However, because the earth is moving as well, this motion sometimes seems erratic. As our planet makes its closest approach to one of these planets, for instance, our own motion relative to the planet will make it seem as though the planet briefly slows down and moves backward. This is called *retrograde motion.*

In recent times, more subtle changes in the night sky have become apparent. For instance, some stars actually move relative to the other stars, though this motion is slow and can only be seen through careful observation and measurements. Still, this motion—called *proper motion*—can cause dramatic changes in the sky

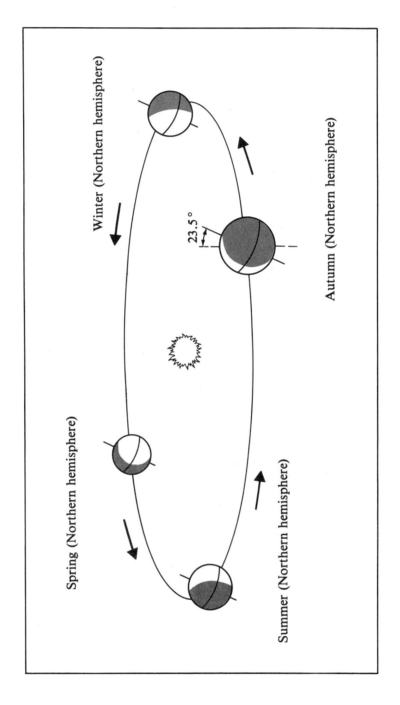

the earth's movement around the sun

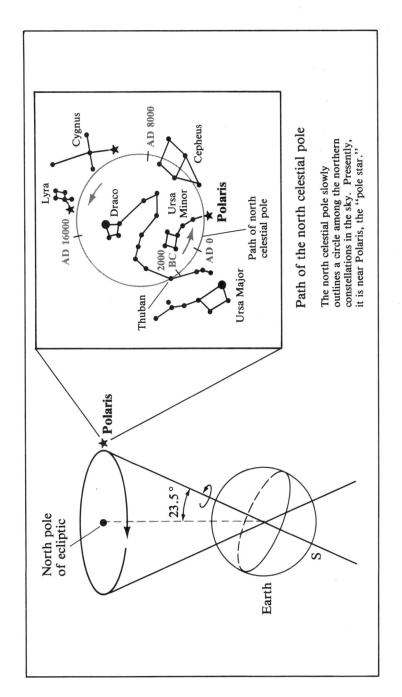

the earth's "wobble" and the precession of the stars

*the Greek scientist Ptolemy and the
Greek mathematician Pythagoras
looking at an astrological book*

over time. Many of the nearer stars are no longer in the same position in the sky today as they were when viewed by the ancient Greeks, for instance.

Not all change in the sky is a matter of motion. Some stars change brightness from day to day or even hour to hour. Such stars are called *variable stars.* In some cases, these are stars that actually produce varying amounts of light. In other cases, they are part of a *multiple-star system,* that is, star systems in which two or more stars are orbiting one another. (When only two stars are orbiting each other, we call them *binary stars.*) Sometimes, one star in a multiple-star system will move in front of another star, as seen from earth. This blocks the light of the first star, and for a time its light seems to fade. Such a situation is called an *eclipsing binary.*

So the night sky is far from constant; to the trained eye, it is a seething hotbed of change. But most of these changes in the sky that we've discussed so far, with the exception of variable stars and the proper motions of stars, were well known to astronomers who lived many thousands of years ago. In fact, these cyclic, predictable changes must have been the first things that ancient people noticed about the sky—after they had noticed the stars themselves.

Indeed, these cyclic changes fascinated the ancient astronomers to a far greater degree than the unchanging motions of the fixed stars. Change, after all, is far more interesting than constancy. An entire pseudoscience called *astrology* developed around these cyclic changes and the effect that they supposedly had on events here on earth. Today, of course, we know that the earthly effects produced by these changes are pretty limited. The apparent change in the sun's path through the sky makes a dramatic difference in the weather, as we saw, and the motion of the moon around the earth produces the tides. Also, certain biological rhythms in plants and animals have apparently synchronized themselves with the motions of the sun and moon, and certainly the cycles of these celestial bodies have psychological effects on human beings. But the ancient astrologers believed that our lives and the course of history itself were determined by the motions of the sun and planets, and in that regard they were wrong. However, anyone who has seen the horoscope column in the daily newspaper knows that this belief has not yet completely died out.

Nonetheless, the ancient astronomers' belief in astrology encouraged them to make a careful study of the motions of the planets. (Prior to modern times, the sun and the moon were considered to be planets just as Venus, Mars, and the others were. The earth, on the other hand, was not regarded as a planet at all.) This detailed study provided a firm background for the science of astronomy, and for this we owe a substantial debt to the astrologers of centuries past.

By the time of the ancient Greeks, whose civilization reached its intellectual peak around 200 B.C., much was known about the way the planets moved through the sky. This knowledge was summed up by the Greek philosopher Aristotle, whose beliefs about the nature of the universe survived, in slightly altered form, for nearly two thousand years.

The view of the universe that came down from Aristotle went something like this: The earth was believed to be at the center of the universe. Everything beyond the earth—the moon, the sun, the other planets, and the stars—was embedded in rotating crystalline spheres that surrounded the earth. The innermost of these spheres contained the moon. The outermost sphere contained the

*the astronomer Hipparchus at the
observatory at Alexandria*

Aristotle

stars. The spheres in between contained the sun and planets.

One of the central tenets of the Aristotelian view of the universe was that everything in the heavens, with the possible exception of the moon, was perfect. (The obviously imperfect moon was that way because it had been polluted by its closeness to the imperfect earth.) The planets, including the moon, moved around the earth in perfect circles, and their motion was unchanging. The sun and the moon were perfect spheres, and the surface of the sun was unblemished.

In the late Middle Ages this view was adopted by the Christian Church, which held that Aristotle's perfect heavens were a testament to the perfection of God. Any views that contradicted the Aristotelian view were considered heretical.

Today we know that the universe is not at all like Aristotle believed it to be. The earth is not at the center of the universe, the planets—with the exception of the moon, which the ancients regarded as a planet—do not orbit the earth, and the heavens are far from perfect.

How did we get from the Aristotelian view of the universe to the modern view? It was a long and rocky path, and it wasn't helped by the fact that the instruments available for studying the skies were quite limited. There were no telescopes until the seventeenth century A.D.; the ancient astronomers had only the naked eye with which to study the motions of the planets. With the naked eye it was impossible to discern the imperfections of the sun, or to obtain sufficient information about the motion of the planets to prove conclusively that they did not orbit the earth. (There was plenty of inconclusive information available, but the ancient astronomers explained it away with some pretty inventive theories.)

the solar system with a comet circling the sun

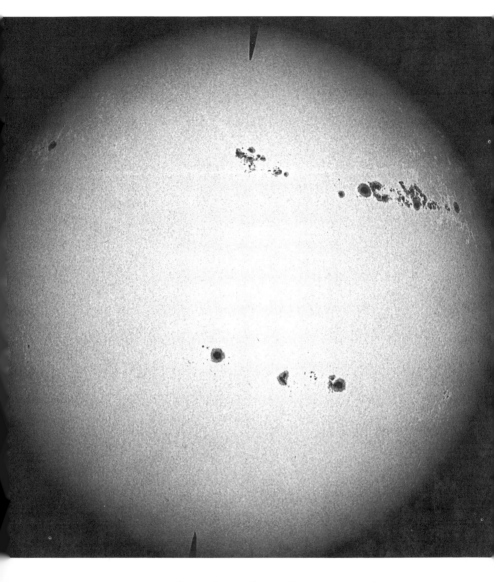

Sunspots, such as those shown here, can only be seen with a telescope. Their discovery proved the sun was not a "perfect" orb, and therefore Aristotle's theories were flawed.

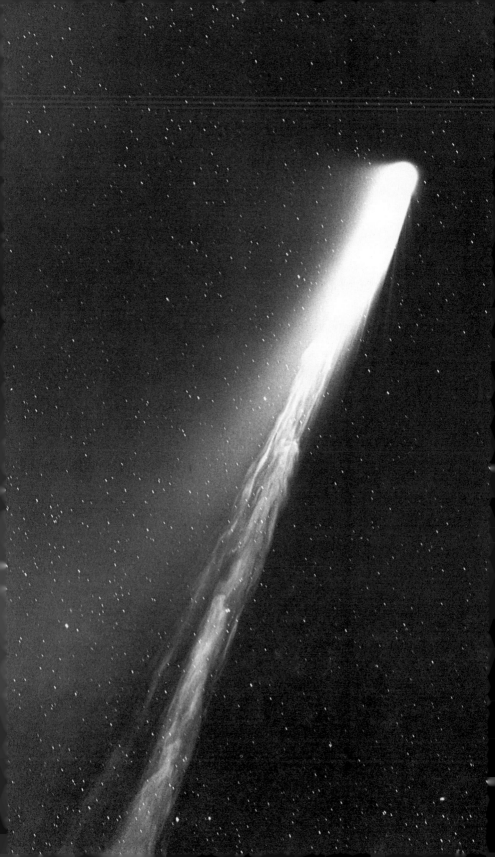

Is there anything visible in the sky, to an astronomer without a telescope, that would have proved that the heavens were imperfect and that the theories of Aristotle were incorrect? To Aristotle, the perfection of the heavens required that they be unchanging, or at least that all changes occur in a predictable, cyclic manner. As we have seen, the sky is full of change, but all the changes described thus far are cyclic and predictable and therefore did not affect the Aristotelian belief in heavenly perfection. Are there any changes in the sky that are not predictable?

Yes. Comets aren't predictable, or at least the ancient astronomers did not know how to predict them. We know today that a comet is a ball of ice and dust left over from the formation of our solar system, though of course the ancient astronomers had no way of knowing this. Billions of comets orbit the sun from a distance of trillions of miles. Every now and then one of them is dislodged from its normal orbit and falls in toward the sun. As it approaches the sun, it moves through the skies of earth for a few weeks, possibly trailing a beautiful and dramatic tail behind it. A few comets, such as the famous Halley's Comet, return to the vicinity of the sun again and again, quite predictably, though it wasn't until much later that anyone noticed this. But others will visit the sun only once, then shoot back out into space for the rest of time—or until something causes them to fall back toward the sun again.

Comets, which are extremely ancient balls of ice and dust, were once considered evil omens.

This very unpredictability made comets a source of fear and confusion in ancient times. Were they an omen of terrible events, people wondered, of the deaths of kings? No, said the astronomer Edmund Halley in the seventeenth century. They were a normal astronomical occurrence, much like the motion of the planets.

Aristotle, however, explained away comets by suggesting that they were a meteorological phenomenon, part of the weather, wisps of fire in the earth's atmosphere. Because they were not part of the perfect heavens, it did not matter that they were not predictable. They did not affect Aristotle's theory.

But there is another type of change in the sky that should have blasted Aristotle's theory into little pieces but for some reason didn't until practically modern times. Perhaps astronomers in Europe, where Aristotle's theories were most accepted, simply refused to notice a phenomenon that contradicted Aristotle so dramatically.

This phenomenon was given the name *nova* in the sixteenth century, from the Latin words *stella nova*, meaning "new star." A nova was a star that appeared in the sky one night where no star had ever been before. And yet a nova was obviously not a star being born, because almost invariably the nova would fade away over the course of weeks, months, or years.

Was a nova, then, a temporary star, a star that only burned for a short time?

No. A nova isn't a temporary star. But it took many thousands of years of observations to learn what novas actually were.

The first nova was observed long before the sixteenth century. In fact, many hundreds of novas have been observed throughout history. A few of these events have been particularly vivid, bright enough to be considered extraordinary events. Many of them were of a special kind of nova that we now call a *supernova*.

When was the first supernova observed? That's difficult to say, since it was almost certainly before the advent of written history.

However, it is almost certain that the brightest supernova in the last ten thousand years was observed about six thousand years ago, at a time when the most advanced civilization on earth lived in the valley between the Tigris and Euphrates rivers. This supernova occurred in the constellation of stars known to us as Vela and is therefore referred to today as the Vela supernova.

What these ancient people saw in the skies would have left a deep impression on humans of any era. At night the Vela supernova would have been brighter than the moon, and in the daytime it would have seemed like a second, dimmer sun.

One expert on the astronomical history of this period, George Michanowsky, has suggested that this supernova had such a profound effect on those who saw it that it led to the invention of writing as a way of recording this incredible event for posterity. Michanowsky contends that the first written word was a symbol for star and was fashioned in the image of the supernova explosion. Later, this symbol was transformed by the ancient Sumerians into another symbol meaning "god." Furthermore, Michanowsky suggests that the ancient Egyptian symbol known as the *ankh,* which resembles an upside-down noose with a bar across it, was an attempt to represent the supernova and its reflection in the ocean. (The ancient Sumerians, who may have invented the ankh symbol, would have seen the supernova just above the horizon, probably over the ocean.)

Almost all scientists and historians believe that

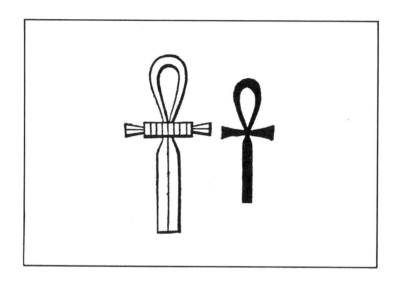

the Egyptian symbol known as the ankh

Michanowsky's theory is a bit farfetched, that it attempts to explain too much with a single astronomical incident. But it does reflect the psychological importance of astronomical events to people on earth and may explain the persistence of the belief in astrology in ancient times, when an event such as a supernova would have been convincing proof that powerful forces resided in the sky.

In the year 1006 a supernova occurred that may have been the brightest since the one in Vela thousands of years earlier. This supernova would have been about one-tenth as bright as the moon. We have records of its appearance from historians in China, though European astronomers don't seem to have noticed it, possibly because it didn't fit into the religious beliefs of the period or was too low in the sky for comfortable viewing.

the supernova of 1054 painted alongside a crescent moon, in sandstone at White Mesa, Arizona

Only half a century later, on July 4, 1054, another supernova, nearly as bright, appeared much higher in the sky. Once again, it was noted by Chinese astronomers, who called it a "guest star," and there are a few historical references to observations of it in Europe. Later, we'll see that it's still possible to view the remains of this supernova today, through a telescope.

In the following centuries, the way in which Europeans regarded the heavens changed dramatically. A Polish astronomer named Nicolaus Copernicus proposed, in the sixteenth century, the essentially modern idea that the earth was not at the center of the universe but that it and the other planets revolved around the sun. Although largely ignored at first, Copernicus's theory eventually began to find adherents among astronomers and other learned people. However, the Christian Church maintained that the Copernican view of the universe was wrong and the Aristotelian view was correct. It took powerful evidence to shake the church's and the public's opposition.

One piece of evidence against the Aristotelian view and its thesis of heavenly perfection appeared in the sky in 1572. In that year a supernova was observed by one of the greatest astronomers of all time, Tycho Brahe of Denmark.

Tycho (as he is known to posterity) did not have a telescope, but he nonetheless studied the supernova in great detail, recording how bright it became (brighter than Venus, the brightest of the planets) and how long it remained visible (485 days). So considerable were Tycho's studies of the supernova that he wrote a book about it, entitled *De Stella Nova* ("About the New Star"). It was the title of this book that gave the phenomenon

the Polish astronomer Nicolaus Copernicus

the Danish astronomer Tycho Brahe

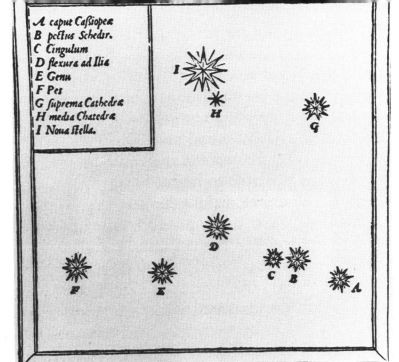

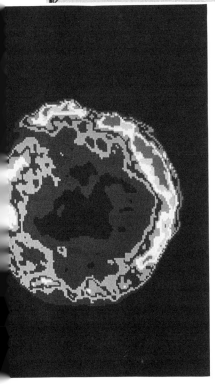

Above: *Tycho's map showing the supernova of 1572 in the constellation Cassiopeia. The star marked "I" on the upper left side is the supernova. The Latin labels, used by Tycho, identify the stars by their positions in the mythological "lady in the chair," which the constellation is supposed to represent.* Below: *a false color radio image of the remnants of Tycho's supernova.*

the German astronomer Johannes Kepler

its name: nova. (The term supernova only came into existence in the twentieth century.)

The book made Tycho's reputation—the nova of 1572 has, in fact, become known as Tycho's supernova—but it was the brilliance of his later observations that assured his reputation as one of the all-time greats in the field. Though he was not a believer in the theories of Copernicus, his observations of the nova of 1572 did much to undermine the belief in Aristotle's perfect heavens, which, in turn, strengthened the general belief in the theories of Copernicus.

In 1604 another supernova, not as bright as the one observed by Tycho but still very bright indeed, flared into view. This one was observed by the great German astronomer Johannes Kepler, who had been Tycho's apprentice some years earlier. Unlike his former master, Kepler was a believer in the theories of Copernicus, and his observations of the nova also helped to chip away at the tyranny of the Aristotelian universe.

It would seem that supernovas were coming hard and fast around the year 1600; the one observed by Kepler came only thirty years after the one observed by Tycho. The invention of the telescope, the greatest of all instruments for observing the sky, came only a few years after Kepler's supernova. It would seem that the time would have been ripe for yet another supernova to appear in the sky, for another generation of astronomers to observe with their new instrument.

But it was not to be. Centuries passed and no nearby supernovas—that is, supernovas that would have been visible to the naked eye, as Tycho's and Kepler's were—appeared in the sky. Now that astronomers had the ability to observe such events in detail, they no longer had such events to observe.

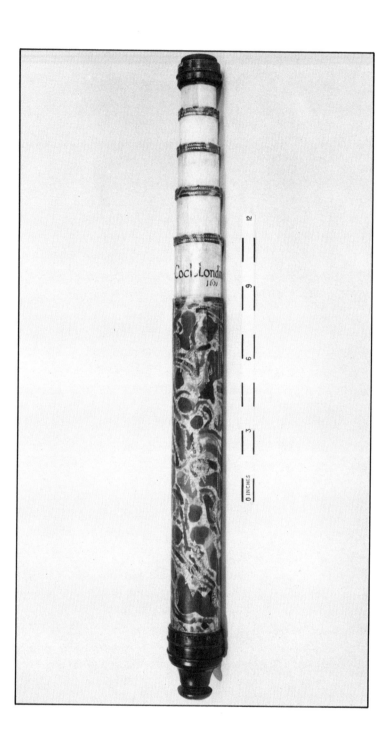

This was a pity, because by the beginning of the twentieth century, science had reached a point in its understanding of the universe where it was reasonable to ask just what a supernova was. What caused this sudden outburst of light in the sky?

To answer this question, it was necessary instead to study the activities of ordinary stars, to ask the more basic question: What is a star and what makes it shine?

an early telescope

2

HOW I WONDER WHAT YOU ARE

The ancient astronomers had no idea what made the stars shine. If they thought about the matter at all, they probably assumed that they were fires in the sky, similar in nature to fires on earth. But what was fueling these fires? Wood, coal, oil? The ancients were in no position to say, though surely they had many theories.

Aristotle believed that objects in the heavens were made out of a different substance than objects on earth. He called this substance *ether* and assumed that it was in the nature of ether to glow, as the stars and planets glowed.

As belief in Aristotle's theories crumbled, however, and belief in Copernicus's theories grew, it became obvious that such simplistic explanations could not suffice. The stars were not made of some sort of magic substance. Why, then, did they glow more brightly than any fire ever known on this planet?

*the sun during a
total solar eclipse*

Actually, most of the stars don't seem terribly bright at all, but today we know that there is one star that astronomers on earth can study close at hand, and it is bright indeed. That star is our sun.

Could the sun be a fire like the fires that we kindle on earth from wood or coal, only much larger? As recently as two hundred years ago that explanation seemed feasible, but by the nineteenth century it was obvious that it suffered from a very large flaw: the sun had been burning for a long, long time. An ordinary fire—even an ordinary fire the size of the sun—would burn out after only a few thousand years at best. Although this might encompass all of recorded history with a little room to spare, there was considerable evidence that the earth, and therefore the sun, had been around for quite a while before human beings appeared on the planet.

Geologists, in fact, had found evidence that the earth was hundreds of millions of years old, at the very least. Paleontologists had uncovered the remains of animals, such as the dinosaurs, that appeared to have lived millions of years ago. Charles Darwin had postulated that *all* living creatures had evolved to their present state through a gradual process called natural selection that would have taken millions of years. Everything, it seemed, pointed to the fact that the earth had been around for a period of at least a hundred million years. So the sun could not be an ordinary fire.

What was it, then? What kind of fire could burn for such a long period of time?

In the mid-nineteenth century, the German physicist Ludwig von Helmholtz took this problem in hand and tried to wrestle from it a solution. He failed, but only because the science of the nineteenth century was not quite ready to provide the answer. Nonetheless, von

the German physicist Ludwig von Helmholtz

Helmholtz's proposed solution provided valuable insights into the processes at work in an ordinary star—and in the very extraordinary interior of a supernova.

Von Helmholtz suggested that it was gravity that kept the sun hot. Gravity, as you may be aware, is the force of attraction produced by all objects in the universe. The planets are held in orbit around the sun by the sun's gravity just as we are held to the earth's surface by the earth's gravity.

Even the individual atoms of which matter is made produce gravity; indeed, the gravity of an object is simply the sum total of the gravity produced by the atoms of which it is made. Although in the nineteenth century atoms were believed to be hard, indivisible little nuggets of substance, we know today that they are made up of still smaller particles—and, yes, these smaller particles also produce gravity.

The sun, of course, is made of atoms, as is almost everything else. Because the interior of the sun behaves in some ways like a liquid or a gas, the atoms of which it is made are free to move around relative to one another.

Von Helmholtz knew that the atoms in the sun produced gravity, and that they would therefore attract one another. This could be causing the sun to contract slowly under its own "weight," as the atoms draw closer to one another under this constant attraction. And this could cause the sun to become hot.

How could this contraction produce heat? Von Helmholtz knew that heat was nothing more than the agitated motion of particles, atoms, molecules, or any entity. As the atoms in the sun contracted and drew together, they would agitate one another and produce still more of this motion—that is, still more heat.

Von Helmholtz calculated the amount by which the sun would need to contract each year in order to produce the observed heat. It turned out to be a very small amount of contraction, too small to be noticed by the observational methods available at the time.

There was a catch, however, a rather obvious catch. If the sun were contracting—that is, growing smaller—it must have once been considerably larger than it is today. In fact, it would once have been so large that it would have encompassed the orbit of the earth! How long ago would the earth have been inside the sun, if von Helmholtz's theory was correct? About 25 million years, he calculated.

Alas, that was not enough time. The geologists said that the earth had been around even longer than that. In fact, today we know that the earth has existed for 4.5 *billion* years. Gravitational contraction simply couldn't keep the sun burning for that long, unless it had recently been very, *very* large. But if it had been so large that recently, the earth could not possibly have existed; it would have been a molten ball swallowed up in the sun's interior. Von Helmholtz had no answer to this objection, and so he retired in defeat.

We shouldn't be too harsh on von Helmholtz, however. He was born in the wrong century. He was unable to provide an answer to the conundrum of the sun because the process that provides the sun's energy wasn't known until the twentieth century.

In the year 1905, Albert Einstein published the famous paper that is known to history as the "Special Theory of Relativity." Among the many amazing hypotheses put forth by this paper was one that suggested that mat-

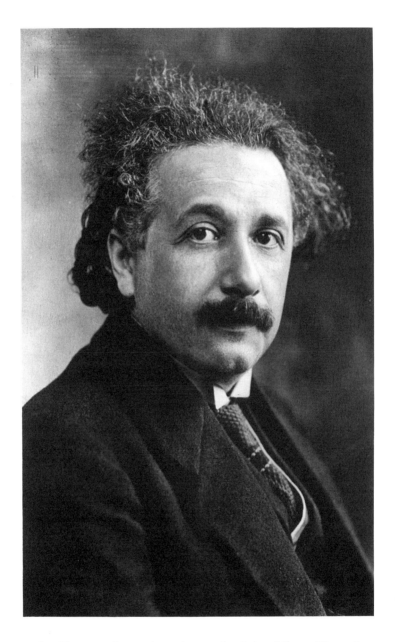

the German-born American physicist Albert Einstein

ter and energy were interchangeable, that under the proper circumstances one could become the other. Einstein even provided an equation that could be used to calculate the precise amount of energy that would be released if a piece of matter were converted into pure energy: $E = MC^2$.

In English, that means that the amount of energy (E) bound up in a given mass (M) of matter is equal to the amount of matter multiplied by the speed of light (C) squared. The speed of light is a very large number (186,000 miles per second, or 300,000 kmps) and when squared—that is, multiplied by itself—it produces a much larger number. Thus, the amount of energy equal to a given amount of mass is quite impressively large.

Ordinary burning may not release enough energy to keep the sun burning for the billions of years that geologists tell us the earth has existed. But if the matter in the sun could somehow be converted to its energy equivalent, according to Einstein's equation, it would provide enough energy to keep the sun bright for billions of years and then some.

Is there a process by which all the matter in the sun could be converted directly into energy? No, not as far as we know. However, it isn't necessary that *all* of the matter in the sun be converted to energy. A fraction of it would be sufficient to keep it burning. And there *is* a process that could convert that fraction into energy. In fact, there is more than one process that would do the trick.

Perhaps the greatest scientific revolution of this century came almost immediately after the year 1900, when physicists began to understand that the atom has an internal structure—that is, that the atom, far from being an indivisible nugget, is made up of still smaller particles and that these particles are put together in a certain way.

A typical atom is made up of three different kinds of particles: *electrons, protons,* and *neutrons.* (According to current theory, some of these particles are in turn made up of smaller particles, but that need not concern us here.) Each of these particles has its own unique personality. Electrons are extremely small, while protons and neutrons are relatively large (though still much smaller than complete atoms, which are themselves minuscule by human standards). An electron has a negative electrical charge while the proton has a positive charge. The neutron has no electrical charge at all; as its name implies, it is neutral.

The neutrons and protons together form the tiny hard core of the atom called the *nucleus.* The electrons inhabit a relatively large shell around this nucleus. In large atoms—that is, atoms larger than those that form the most common elements, hydrogen and helium—there are several of these shells of electrons, nested inside one another like Oriental gift boxes. Powerful forces, much more powerful than the force of gravity, work inside the atom to maintain this structure. Without these forces, the structure that we call an atom could not exist. The universe would be a sea of tiny particles, without stars, planets, or life.

Although the forces that hold the atomic structure together are powerful, it is nonetheless possible for atoms to become rearranged, knocked apart, and put together in new ways. The simplest rearrangement of atoms concerns the electrons. An atom can lose electrons or gain new ones. Under certain conditions, such as inside the sun, atoms commonly exist with no electron shells at all.

a helium atom

Rearrangements of the nucleus of an atom, of the protons and neutrons at the core of an atom, are far less common, though under certain circumstances they happen spontaneously. Certain types of uranium and plutonium atoms, for instance, have nuclei that are not stable. Every now and then, quite unpredictably, one will break apart and form a pair of brand-new nuclei, which in turn will become the cores of new atoms.

When this occurs, something rather peculiar takes place. A tiny part of the matter in the atom turns into energy, according to Einstein's equations. If enough plutonium or uranium atoms broke apart (or *fissioned,* as physicists call it) at the same time, the amount of energy released would be huge. But if the breakdown of a uranium or plutonium atom is unpredictable, how can we cause a large number of such atoms to break down all at once?

It is possible to coax the fission process along by bombarding unstable nuclei with fast-moving particles. When uranium or plutonium nuclei fission, they send high-speed protons zooming outward. These speeding protons can in turn collide with other unstable nuclei, causing them to fission as well. Since these nuclei also emit high-speed protons, they can cause still other nuclei to fission, and so forth. This cascade of fissioning nuclei is called a *chain reaction*. To create a chain reaction, we need merely bring a large number of such atoms together. The amount of an unstable element that must be brought together to create a chain reaction is called the *critical mass*.

This is the principle behind both the atomic bomb and nuclear reactors. If we bring a critical mass of unstable plutonium or uranium atoms together quickly enough, the chain reaction will proceed so rapidly that a tremendous explosion will take place. If we bring the critical mass together somewhat more slowly, huge amounts of heat (but no explosion) will be produced. This heat can be used to create steam that can be used to turn a turbine.

Fission is a much better method of generating energy than ordinary burning. Could it be the source of the sun's energy?

Not really. There simply aren't enough unstable atoms in the sun to reach critical mass. In fact, there are virtually none. Almost all of the atoms in the sun are of the elements hydrogen and helium, and they are quite stable.

A mushroom cloud forms over a small uninhabited island in the Pacific after the explosion of an atomic bomb.

However, there is a second type of rearrangement that can take place within atoms, and it also produces energy. This type of rearrangement is called *fusion*.

In a fusion reaction, atoms come together at high speeds, knock one another apart, then reform into atoms that are larger and heavier than the original atoms were. Actually, fusion reactions are even more complicated than this description implies, usually involving quite a few atoms, some of which are changed by the reaction and some of which are not. But you may imagine a fusion reaction as one in which two atoms fuse together to form one larger atom.

As with fission, a small quantity of matter is converted into pure energy during the fusion reaction. And, in fact, the quantity converted during fusion is larger than the quantity converted during fission, making fusion a much more efficient source of energy, one of the most efficient known. It is fusion that makes the hydrogen bomb explode. Alas, no way has yet been found of using fusion to efficiently power a nuclear reactor. Some scientists believe that such a nuclear reactor, if it could be built, would answer many of the world's energy problems.

In the late 1930s, German-born American physicist Hans Bethe proved that fusion is the source of the sun's heat, which had been suspected by earlier physicists. Specifically, the sun gets its energy from *hydrogen fusion,* the fusion of hydrogen atoms to create helium atoms. Not only did this explain why the sun could burn for billions of years, but it helped to explain the presence of hydrogen and helium in the sun. The hydrogen was the fuel that the sun "burned," and the helium was the "ash" left over from the burning.

the German-born American physicist Hans Bethe

So, by the middle of this century, scientists understood the essential process that made the sun, and presumably most other stars, shine. But how do supernovas fit into this picture? If a supernova is a kind of star, does it shine by fusing hydrogen atoms into helium as our sun does? If so, why does it shine so briefly? What makes it flare up suddenly and then disappear?

The answer is that a supernova is *not* a normal star; it is something that *happens* to certain stars. To better understand how supernovas work, it's necessary that we think of stars as having lifetimes, just as human beings do, with a birth, a middle age, a death—and even a life after death.

3
LIKE A DIAMOND IN THE SKY

Space isn't empty. Although we think of outer space as a vacuum, an area with no material in it whatsoever, it is actually filled with tiny particles, atoms of gas and dust. We refer to this cloud of material between the stars as the *interstellar medium*.

To be honest, the interstellar medium is pretty thin. On earth, we would be forgiven for mistaking it for a hard vacuum, since it is thinner than the best vacuums ever produced on this planet. But astronomers who study distant stars and galaxies must take this interstellar medium into account, because it discolors the light that reaches their telescopes, as though they were looking at the stars through a slightly dingy film, a waxy yellow buildup in space.

As thin as it is, there are places where the interstellar medium becomes relatively thick (though if it were on earth we still might mistake it for a vacuum). Viewed through telescopes—and, in a few cases, with the naked

the Horsehead Nebula, filling the void of space with its cloud of gas and dust

eye—these areas look like clouds hanging between the stars. And, in fact, that is what they are: clouds of gas and dust. Mostly, these clouds are made up of hydrogen and helium, the primary constituents of the universe, but they usually contain other "impurities" as well, among which are numbered most of the elements in the periodic table.

If something happens to one of these clouds—if, say, it is struck by the shock wave from a nearby explosion, of a sort that we will consider in a moment—it can contract, just as von Helmholtz believed that the sun was contracting, with the atoms in the cloud exercising their gravitational attraction on one another. As the cloud grows smaller, it also grows hotter, from the same gravitational energy that von Helmholtz believed to be the source of the sun's heat. After a few million years of contracting, the cloud would look a lot like a star.

At a certain point, when the hot atoms in the cloud began moving rapidly enough to break apart under the force of their own collisions, hydrogen fusion would begin in the cloud. This fusion, in turn, would provide enough heat to *keep* the hydrogen fusion going. The pressure of the fusion inside the cloud would prevent the gravity of the atoms from collapsing the cloud still further.

And at that point the cloud of gas and dust would be a star, just as our own sun is. In fact, this is almost certainly how our sun began its illustrious career about 5 billion years ago. Once the hydrogen fusion balances the gravitational collapse of the cloud, the star becomes quite stable. It remains in this state for a long period of time.

Just how long a period of time depends on how large the star is. Very large stars are much brighter and hotter than smaller stars and thus burn up their hydrogen fuel very quickly; such stars have lifetimes measurable in the millions of years. Very small stars, on the other hand, are relatively cool and can keep burning for hundreds of billions of years. Our sun falls somewhere in between. It has a predicted hydrogen-burning lifetime of about 10 billion years, approximately 5 billion of which are already over.

What will happen at the end of that 10 billion years? Will the sun run out of hydrogen and be snuffed out like a candle? Not exactly. As the sun burns hydrogen, it produces helium atoms. Eventually, this helium ash will choke the central furnace of the sun and hydrogen fusion will simply cease, even though 90 percent of the sun's hydrogen will remain unburned.

Since there will no longer be any fusion reactions to balance out the tendency of the sun to collapse gravitationally, the sun will begin to collapse again—but only for a moment. Gravitational energy will produce such incredibly high temperatures that helium fusion will begin in the sun's core—that is, atoms of helium will begin fusing together to form atoms of carbon.

So hot will the center of the sun become at this point that the star will swell into a *red giant,* a star so large that it will swallow up the earth (much as von Helmholtz incorrectly suspected it might have done in the distant past). This, in fact, will be the eventual fate of most stars, since most stars will ultimately run out of hydrogen and begin fusing helium. A few of the stars visible to the naked eye are red giants, the best known being Betelgeuse.

Red giants are fairly rare, though, because no star can remain in this state for long. Helium burns quickly, and the helium supply of a red giant is not terribly large

Orion's Belt consists of the three stars in the center of the photograph. Betelgeuse is above and to the right.

to begin with. After less than a billion years a star the size of our sun will generate enough carbon ash to choke off the helium fusion reactions. Then it will begin to collapse yet again, like a punctured balloon. It will never generate enough heat to fuse carbon atoms into still larger atoms, though such a reaction is possible. It will only stop collapsing when all of its atoms are jammed up tightly against one another, in a dense ball.

The atoms in the collapsed sun will be tightly packed indeed, much more tightly packed than atoms ever become here on earth. Such tightly packed matter is called *degenerate matter,* because it has degenerated far beyond the point that matter is allowed to under ordinary circumstances; on earth, a teaspoon of this matter would weigh a ton. The force that stops the collapse of the star is called *electron degeneracy pressure;* in effect, the electrons in the star refuse to approach one another any more closely.

The ball of such matter left over from the collapse of the sun will be a *white dwarf,* a small but dense, collapsed star. The structure of a white dwarf resembles an extremely dense crystal, rather like a giant diamond. It will glow from the residual heat of fusion reactions and gravitational energy but will eventually cool off to become a *black dwarf.*

Astronomers have observed white dwarfs orbiting the larger stars in multiple-star systems. Usually, the white dwarfs themselves are too small and far away to be seen, but the effect of their gravity on the other stars in the system can be dramatic. The first white dwarf, for instance, was discovered by the "wobble" that it created in its companion star, Sirius. No observation of a *black* dwarf has ever been confirmed, probably because they are extremely hard to see.

*three exposures of Sirius, showing
the companion, a white dwarf*

The previous paragraphs describe the life cycle of a typical star, like our sun. But not all stars are typical. In particular, very massive stars experience a somewhat different sequence of events near the end of their lives.

A star like our sun burns hydrogen during its "normal" lifetime, then burns helium during its brief existence as a red giant. When the red giant's helium core fills with carbon ash, the red giant collapses, and no more reactions take place to keep it burning.

This is not the case with a very massive star. When the very massive star stops burning helium, it generates so much heat through gravitational collapse that it begins to burn carbon, fusing the carbon atoms into neon and magnesium atoms. As the various atoms in the star's core are converted into heavier atoms, the core eventually fills up with iron atoms.

And there the process ends. The iron atoms will not fuse into larger atoms. The energy required is just too great. Filling the core of a star with iron atoms is like throwing water on a fire; it can no longer keep burning.

Although it would seem that the ability to burn all of these various types of atoms would keep the massive star alive for a very long time, this isn't actually so. If helium burns quickly, these heavier elements burn more rapidly still. Core after core of fusion ash form quickly within the star, with each core forming inside the previous core, until the inside of the star resembles a layered onion. And then the iron ash chokes off the fusion reactions altogether.

What then? Does the large star collapse just as a small star does? Yes, at first. But the force of gravity within the star causes it to collapse so rapidly that it cannot simply become a white dwarf. Even the electron

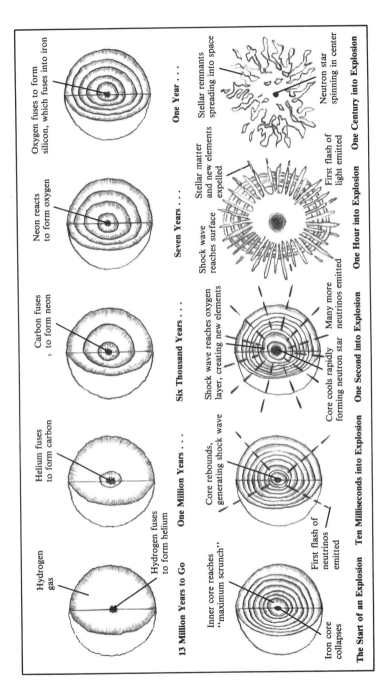

the steps leading to a supernova

degeneracy pressure that stops a smaller star from collapsing into anything denser than a white dwarf does not stop the collapse of the star. Instead, it collapses to a certain size, then stops. The tightly packed atoms inside the star begin to act like a coiled spring, causing the matter in the outer part of the star to shoot back out into space with tremendous force, while the interior of the star continues to collapse.

The result is an explosion of staggering proportions. The incredible energy with which the material of the star is flung outward produces fusion reactions not ordinarily found within a star. For one brief moment the star shines more brightly than 10 billion ordinary stars.

As you may have guessed, this explosion is a supernova—or, at least, it is one type of supernova. (We'll look at some other types of supernovas—and of just plain novas—in the next chapter.) It is almost certainly what Tycho Brahe saw in the skies of 1572, what Johannes Kepler saw in 1604, and what the Chinese astronomers saw in 1006 and 1054. One scientist has estimated that there is one supernova every second in the universe.

Surprising things happen within the supernova. Atoms fuse together with abandon, forming elements not ordinarily created by fusion, including some of the elements heavier than iron. In fact, this is how, current theory tells us, most of the atoms heavier than hydrogen were formed: in the heart of a star. This includes most of the atoms in your own body, the atoms in this book, the atoms in the chair you are sitting on, and so forth. Without the atoms formed in the supernova explosion, most things in the universe made out of elements other than hydrogen and helium—including the planet earth and the beings that live on it—would not be possible.

Actually, many of these elements are formed during earlier fusion processes that take place during the normal evolution of the star, but it is the supernova explosion that flings them out into the universe, where they can become part of the clouds of gas and dust that form stars and planets. Stars that are rich in heavy atoms formed by earlier supernova explosions are referred to by astronomers as *Population I* stars. Stars that are poor in such heavy, supernova-formed atoms are called *Population II* stars. And, in fact, it was probably a supernova that caused the shock wave that triggered the collapse of the star that became our sun, as described in the last chapter.

What about the core of the massive star, the part that kept collapsing while the outer portion of the star was flung outward into space? Does it become a white dwarf as a smaller star does when it collapses?

No. As we saw above, the force with which the star collapses is sufficient to overcome the electron degeneracy pressure that stops an ordinary star from collapsing. The very nuclei of the atoms come into contact with one another. Then a decidedly odd thing happens. The nuclei merge and the electrons and protons come in contact with one another. They join together to become neutrons, and the interior of the star becomes almost a solid mass of neutrons, with perhaps a few protons remaining. It is the *nuclear force,* which under normal circumstances holds the nuclei of atoms together, that keeps the neutrons from coming any closer to one another—and thus prevents the star from collapsing beyond a certain point.

By this point, however, it is much smaller and denser than a white dwarf. It has become, in fact, a *neutron star*—a star made entirely of neutrons. A typical neutron star ends up only 10 or 15 miles (16 or 24 km) in diameter, after collapsing from a star that was many millions of miles in diameter.

Neutron stars were first suggested by the theories of the Indian-born astronomer Subrahmanyan Chandrasekhar in the 1930s. But the astronomical community was not yet ready for such an idea. The great English astronomer Sir Arthur Eddington called it a *reductio ad absurdum*—an idea carried to the point of absurdity—and refused to accept it for as long as he lived.

But time proved Chandrasekhar to have been more farsighted than Eddington. In 1967, an English graduate student named Jocelyn Bell discovered that the radio telescope she was monitoring was picking up unusual signals from space each time it swept a certain portion of the sky, signals that she referred to as "scruff." The most notable characteristic of this scruff was how absolutely clean and even it was, a sequence of radio signals exactly one and one-third seconds apart. It was almost as though she were detecting an interstellar navigation beacon, or some kind of carrier wave being broadcast by an alien race to catch the attention of radio astronomers on distant planets.

She turned the matter over to her teacher, the English astronomer Antony Hewish. For a time the two of them wondered if they had inadvertently contacted an alien race, which they nicknamed the LGM (for Little Green Men). But soon they discovered another source of "scruff" in a completely different part of the sky. Since it seemed unlikely that more than one group of aliens was sending the same signal, they looked instead for a natural source to explain the signal.

When they announced their discovery, the scruff

was soon dubbed a *pulsating star,* or *pulsar* for short, because it emitted such evenly spaced radio pulses. Astronomers around the world looked for an explanation. If the pulses were indeed the result of a natural phenomenon, it must be a phenomenon that repeated itself on an extremely regular basis. The most likely such phenomenon was the rotation of a star or planet, but nobody knew of any large object in the sky that could rotate in one and one-third seconds. (Later, pulsars were discovered that emitted signals at an even faster rate, indicating an even faster rotation.)

But the object emitting the signals must also be hot, because a great deal of energy is necessary to produce the signals that radio astronomers were detecting. It was probably a kind of star, but no type of star known, including a white dwarf, was small enough to rotate at the speeds that a pulsar must be rotating.

Except neutron stars.

However, no one had ever proved that neutron stars existed. In fact, the idea had been largely ignored since Chandrasekhar had first suggested it three decades earlier. It had just seemed too incredible to contemplate. But now the idea was hauled back out of the closet and used to explain the signals of the pulsars.

Basically, the theory that was developed to explain the pulsars tells us that the neutron star spins very rapidly, in some cases rotating many times a second. As it rotates, the lines of magnetic force surrounding the neutron star—the same lines of magnetic force that on earth determine the direction in which a compass points—become entangled and generate a powerful radio signal, which "snaps" out into space like the end of a whip every time the neutron star rotates. This cosmic beacon is picked up by any radio receiver in the direction in which it is snapped. Thus, we can only pick up neutron star signals that are heading in our direction.

Is there any concrete evidence that neutron stars and pulsars are the same thing? Yes. Not long after Bell and Hewish discovered their pulsars, a team of astronomers aimed their radio telescope at a cloud of gas and dust in space known as the Crab Nebula and discovered that there was a pulsar there.

Why is this significant? The Crab Nebula is believed to be the remnant left behind by the supernova explosion observed by Chinese astronomers in the year 1054. The cloud is in the right part of the sky, and the rate at which it is expanding outward would have brought it to its present size in exactly the number of years since the explosion was observed. The pulsar in the Crab Nebula is almost certainly the neutron star left behind by this explosion. Fortunately, it is "aimed" in the proper direction to be heard by radio telescopes on earth.

Neutron stars are not the only kind of remnant left behind by supernova explosions. Very large stars produce such tremendous pressures in their interiors as they collapse that even the nuclear force that holds a neutron star together is overcome. There is no known force that acts on a smaller scale than the nuclear force; hence, there is nothing—as far as we know, anyway—that will

Above: *the Crab Nebula in Taurus.* Below: *an X-ray picture of the Crab Nebula showing the pulsar, which is the bright circular object in the center.*

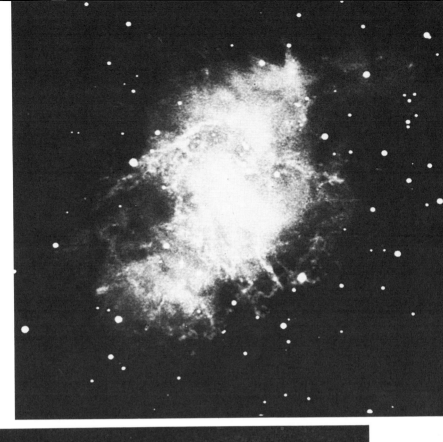

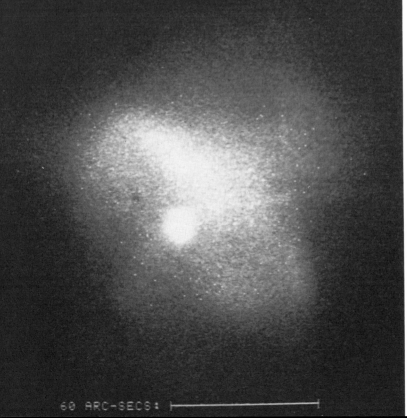

60 ARC-SECS

stop such a star from collapsing. It will simply continue to collapse indefinitely.

This is a rather frightening thought. A star that continues collapsing forever will eventually become infinitely small, a very difficult concept to contemplate. However, physicists are able to theorize about what goes on in the vicinity of such a star. Though they disagree on some points, such as what happens at the center of such an infinitely dense star, they do agree on one striking attribute of such stars: they would produce so much gravity that even light would be unable to escape from them. The still hot and glowing star would appear jet black to an observer. It would become a kind of hole in space, a *black hole*. And that, in fact, is the name by which we refer to such supercollapsed stars.

Scientists tell us that other strange things will happen in the vicinity of a black hole. Time will slow down, so that a person falling into a black hole would experience time passing at a different rate than a person outside the black hole. Some theorists have even suggested that the black hole could be the opening to a passageway that leads to other places in space and even other times. It has even been speculated that black holes could be used as a means of traveling through time, or for traveling great distances across space—assuming that we can find a way to go through the hole without being destroyed by the terrible forces that surround it and without being trapped by its intense gravity.

How remarkable that a single phenomenon—supernovas—could produce two additional phenomena as amazing as neutron stars and black holes. By now, however, you may be wondering how we know so many things about supernovas if we have never had the opportunity to study one up close.

4
THE MANY TYPES OF NOVAS

The last nearby supernova explosion occurred in 1604, only five years before the invention of the telescope. If it had only occurred a few years later it could have been studied in much greater detail than it was, and much more could have been learned about it.

As the science of astronomy has progressed, astronomers have developed more and more precise tools for studying phenomena in outer space. Today it is possible to tell what elements a star is made of simply by studying the light that it produces. It is also possible to study other kinds of radiation released by stars, such as X rays, gamma rays, radio waves, and ultraviolet and infrared light. With these instruments, we could learn a great deal about supernovas, if only we had a nearby one to observe. Yet, for nearly four centuries, no such supernovas have occurred.

This doesn't mean, however, that astronomers have been unable to observe supernovas. In fact, they have

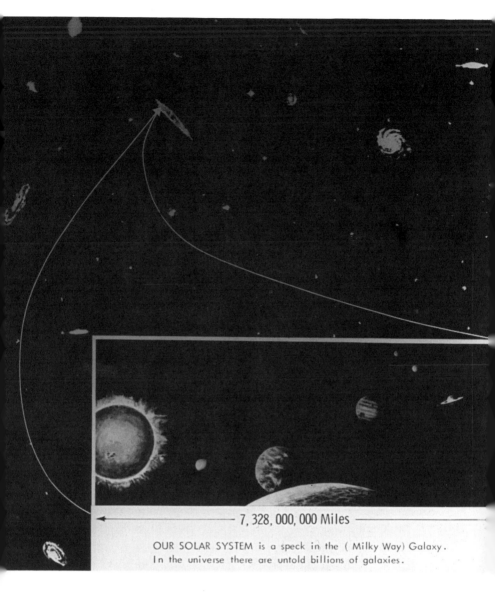

7, 328, 000, 000 Miles

OUR SOLAR SYSTEM is a speck in the (Milky Way) Galaxy. In the universe there are untold billions of galaxies.

*There are millions of galaxies in the
universe visible to our telescopes,
and they come in lots of shapes and sizes.*

observed over six hundred of them. But they have all been very far away.

The supernovas observed by Tycho, Kepler, and the Chinese astronomers all occurred in the Milky Way Galaxy, the vast disc of stars of which our sun is a part. This means that they were all relatively nearby, close enough to observe with the naked eye.

By some calculations, a supernova should occur in the Milky Way Galaxy about once every twenty-five years. Unfortunately, we can only see a small fraction of these, because there are vast clouds of gas and dust that block our view of most stars in the galaxy. Still, a supernova should occur in the one-tenth of the galaxy that we can see about once every two hundred and fifty years. It is sheer bad luck that no supernovas have occurred in our small section of the galaxy in nearly four hundred years, since the invention of the telescope.

However, we can also observe other galaxies through our telescopes, and supernovas occur in those galaxies as well. Most of these other galaxies are quite far away, but a supernova produces a tremendous amount of light. In fact, a typical supernova produces as much light as all of the other stars in its galaxy put together. Thus, if a galaxy is close enough to be viewed through our telescopes, it is close enough for us to see supernovas in it, assuming that clouds of gas within that galaxy don't get in the way. And there are millions of galaxies that are visible to our telescopes.

The pioneers in this research were the Swiss-born astronomer Fritz Zwicky and German-born astronomer Walter Baade, in the 1930s. In fact, it was Zwicky and Baade who coined the term supernova, to distinguish this phenomenon from a kind of variable star that has come to be known as a nova. (You'll recall, however, that the term nova was coined by Tycho in reference to a star that was a supernova.) We'll have more to say about novas later in this chapter.

Zwicky and Baade developed much of the theory of supernovas that we discussed in the last chapter, though of course this theory has been polished and added to in the intervening years. (Zwicky and Baade, for instance, were not aware of black holes.) Just as Chandrasekhar's related theory of white dwarf stars caused considerable controversy when it was proposed at about the same time, Zwicky and Baade met with outspoken opposition for their theory of supernovas—much of it from the same people who opposed Chandrasekhar. Sir Arthur Eddington, for instance, refused to believe that a star could end its life cycle as anything but a white dwarf.

However, the observations of supernovas in other galaxies did much to support Zwicky's and Baade's beliefs. Up until the 1920s, scientists weren't even sure that other galaxies existed; according to some theories the galaxies that were observed through telescopes were mere clouds of gas, and the Milky Way represented the whole of the universe. The observations of supernovas helped to abolish this misconception, since it is unlikely that faint new stars would appear spontaneously in gas clouds on such a regular basis.

To date, more than six hundred supernovas have been viewed in other galaxies, an impressive number. Zwicky and the team of astronomers with which he worked found almost half of these over the course of several decades' work. Currently, between twenty and twenty-five new supernovas are spotted by astronomers every year.

We can also observe the debris scattered into space by supernova explosions from the past. As we have already seen, the Crab Nebula is the cloud of gas and dust left over from the supernova of 1054; this cloud is still expanding outward rapidly from the center of the explosion and is being heated by energy from the neutron star at its center. A small cloud has also been detected at the place in the sky where Kepler saw his supernova, and there are unusual radio signals emanating from the area of Tycho's nova.

But perhaps the most interesting suspected supernova remnant is the so-called Gum Nebula, named for the Australian astronomer Colin S. Gum. The Gum Nebula is very thin and spread out, but it is so close to the earth that it covers nearly one-sixteenth of the sky. It is not visible to the naked eye, but it can be studied through telescopes in the Southern Hemisphere. Apparently it was the site of a supernova explosion about thirty thousand years ago.

Confusing the issue of supernova observations is the fact that supernovas of the sort we described in the last chapter are not the only sudden explosions of light visible in the sky. There are, for instance, the novas that we mentioned earlier. And there is more than one type of supernova.

The phenomenon that we refer to as a nova looks much like a supernova—that is, it is a sudden flare-up of light in the sky that gives the impression of being a brand-new star, hence the name nova. Of course, it is not a brand-new star. Like a supernova, it is actually a flare-up of an older star.

Novas are much dimmer than supernovas, by several orders of magnitude. However, that is not always obvious when observing such events in the sky, since a nearby nova may appear to be brighter than a very distant supernova, the way that a light bulb next to your bed may appear brighter than the streetlamp outside your window, even though the streetlamp would be considerably brighter if viewed close up. But it was only about half a century ago, when astronomers gained the necessary tools for measuring the distance to stars, that it became possible to tell the difference.

Furthermore, a nova can occur more than once, while a supernova is a one-time-only event. This is why a nova is considered to be a type of variable star, like the ones mentioned in the first chapter.

Is a nova an exploding star, like a supernova? Sort of, but not exactly. The best current guess is that a nova is part of a double-star system, where a white dwarf star is orbiting around a much larger star. If the stars in such a system are very close together, the gravity of the white dwarf will pull hot material off of the larger star, and over a period of time it will build up on the white dwarf's surface. When a certain mass of material has built up on the white dwarf, the gravitational energy produced by that material will cause it to become hot so rapidly that it will explode and blow away most of the material into space. Because this process can start up again immediately after the explosion, the nova can explode again and again.

There are also two types of supernovas, and one of them is remarkably similar to an ordinary nova. They are called, rather unimaginatively, *Type I* and *Type II* supernovas. The kind of supernova that we described in the last chapter—an old star that explodes in the final stages of its evolution—is a Type II supernova. A Type I supernova, on the other hand, occurs in almost exactly the same way as the nova described in the previous paragraph, with a white dwarf drawing material off from a companion star. The difference is that the white dwarf is extremely large, probably at the upper limit of the white-dwarf-size range. Because of its tremendous mass, the material added to the white dwarf will cause an extraordinarily large explosion, which will leave nothing behind. Thus, unlike a nova, a Type I supernova can never recur. Neither does it produce a neutron star or black hole.

Astronomers have been studying supernovas in other galaxies for more than a century now, but there is much that they are unable to learn about these exploding stars when viewed at such a distance. One of the things that astronomers have wanted to study about supernovas is the way in which they produce *neutrinos*.

What is a neutrino? It is a tiny particle a lot like the ones that make up the atom—the electron, the proton, and the neutron—but with some important differences.

For one thing, it is nearly invisible. That is, it cannot be detected with the standard instruments that physicists use to detect subatomic particles, particles smaller than atoms. (All particles of this size, including the atom itself, are of course invisible to our eyes, but physicists long ago devised methods of "seeing" these particles with instruments other than the eye.)

We only know of the neutrino because of the inspired guesswork of an Austrian-born physicist named Wolfgang Pauli. In the 1930s, Pauli was studying the disintegration of unstable atomic nuclei, as we described in Chapter Two. Pauli noticed that a certain amount of energy got "lost" during this process, or at least he was

the Austrian-born American physicist Wolfgang Pauli

unable to account for it. His inspired guess was that a particle he called the neutrino was created during the disintegration of an unstable nucleus and that this particle "stole" the energy as it rocketed out of the atom. Working entirely with theoretical equations, Pauli deduced what the neutrino must be like.

What is it like? It's very odd indeed. For one thing, it has no mass, according to Pauli. (Mass is the property that gives objects and particles weight when they are in a gravitational field, such as the gravitational field of earth. In effect, Pauli was saying that the neutrino weighed nothing at all.) It's also very reluctant to interact with other particles. A neutrino can pass through a brick wall trillions of miles thick without even slowing down. This latter feature explains why physicists had not previously observed neutrinos with standard particle-detecting equipment: the neutrinos rarely interacted with the equipment!

Not only are neutrinos produced during the disintegration of unstable atoms, but they are also produced during fusion. Thus, our sun should be producing large numbers of neutrinos at this very moment. Billions of neutrinos are passing through this book—and through your body—as you read.

In order to detect these neutrinos from the sun, scientists have built neutrino telescopes and placed them at various locations around the world. How can you build a telescope to detect something that doesn't interact with detecting equipment? Fortunately, a neutrino will occasionally interact with another particle, almost by accident. This is an extremely rare event, but if we build a large enough detector eventually a neutrino will interact with it.

A neutrino telescope doesn't look like any telescope you've ever seen. One type of neutrino detector is an extremely large tank filled with chemicals similar to those used to clean linoleum floors. These tanks are placed far beneath the surface of the earth in underground mines, where other types of particles never reach; neutrinos, of course, are undeterred by having miles of dirt placed between them and the telescope. When a neutrino interacts with a particle of the liquid in the tank, it alters that particle in a way that can be detected by conventional detectors. Instruments attached to the vat keep a record of all such "neutrino events" within the vat. Since we know—or, at least, we think we know—what percentage of neutrinos will interact with the liquid, we can calculate how many neutrinos actually pass through the vat based on how many are detected. Neutrinos can also be observed by devices designed to detect the possible decay of the proton into smaller particles.

When scientists first built such telescopes they were surprised to detect much fewer neutrinos emanating from the sun than theory had predicted. Was something wrong with the neutrino telescopes? Was something wrong with the theory of hydrogen fusion? Or—horror of horrors!—was something wrong with our sun?

More likely, it is our *understanding* of the sun that is at fault. But in what way? Scientists have devised many clever theories to explain the absence of solar neutrinos, but none has yet emerged as the most likely explanation. Perhaps someday soon a new theory will explain why the sun is producing fewer neutrinos than we think it ought to.

Supernovas also produce neutrinos. But current supernova theory holds that the supernova produces an unusually large burst of neutrinos just before the explo-

An underground proton decay detector, which is basically a tank of ultrapure water with photomultiplier tubes sticking into it, is used to detect neutrinos.

sion. In fact, most of the energy released by a supernova explosion goes into the production of neutrinos; only a relatively small amount emerges in the form of light, though even that "small" amount of energy is enough to make a supernova far and away the brightest object in a typical galaxy. Since this burst of neutrinos is crucial to the theory, astronomers would like to study a supernova with neutrino telescopes. But supernovas in distant galaxies are much too far away for their neutrinos to be detected with any reliability.

Observation of a supernova with a neutrino telescope would also tell us something about the neutrino itself. It would help us determine whether or not it has mass.

Wolfgang Pauli said that neutrinos had no mass, and for years physicists assumed this to be correct. But in the 1970s a few scientists began to question this assumption. Perhaps, they said, neutrinos had an extremely *tiny* amount of mass, too little to be detected by our instruments.

This would seem a rather unimportant argument, until you consider that neutrinos are far and away the most abundant particles in the universe. Even if the mass of a neutrino were the tiniest fraction of the mass of a particle like a proton or even an electron, all of the neutrinos in the universe would still outweigh all of the other particles in the universe by a considerable margin. And the gravity produced by all those neutrinos could have important consequences for the long-term fate of the universe, as we will see later.

How could observations of a supernova tell us how much neutrinos weigh? If neutrinos have no mass, then theory tells us that they must travel at the speed of light, which is very fast indeed. This means that the neutrinos from a supernova will travel from the exploding star to our telescopes just as fast as the light from the explosion does. Since the outburst of neutrinos is able to "escape" the star before the light from the explosion does, we should actually "see" the neutrinos shortly before we see the explosion itself. Thus, the timing of the two events—the detection of the neutrinos and the detection of light from the explosion—could solve the riddle of the neutrino's mass.

Needless to say, astronomers have been frustrated by the dearth of nearby supernovas since Kepler observed the nova of 1604. Today we have a vast array of instruments with which a supernova could be studied in great detail, if one would only occur close enough to earth. (Of course, we don't want a supernova to occur *too* close to earth, since the eradiation from such an explosion could be deadly to any creature living within trillions of miles of it!)

Year after year astronomers have scanned the skies hoping for such an event to occur. For a while it had begun to seem like the watched pot of the heavens was never going to boil. . . .

Until Supernova 1987A blazed out in the Large Magellanic Cloud.

5

HOT TIMES IN THE MAGELLANIC CLOUD

There have been no supernovas visible in the Milky Way Galaxy since Kepler's nova of 1604. That statement is as true today—or, at least, at the time of this book's writing—as it has been since the last flickering light of the 1604 supernova disappeared from the sky. The small portion of our galaxy that can be viewed through our telescopes has been frustratingly supernova-free for the better part of four centuries.

Supernova 1987A, which Ian Shelton photographed on that windy February night at the Las Campanas observatory, was not in the Milky Way. Rather, it was in the Large Magellanic Cloud, or LMC, a small, irregularly shaped companion galaxy to our own.

In several ways, this is an advantage. For one thing, the LMC is not terribly far away in astronomical terms—roughly 170,000 light-years, the distance that light travels in 170,000 years. This means that the light we see from the LMC today actually began its journey 170,000

*the Large Magellanic Cloud,
about 170,000 light-years away*

years ago. Supernova 1987A, then, took place around 168,000 B.C., give or take a few thousand years, though we are just seeing it now.

This may sound like a lot of light-years, but it isn't really. The Milky Way Galaxy is roughly 100,000 light-years in diameter, so the LMC is no more than twice as far away from us as some parts of our own galaxy, and it is closer to us than any other galaxy. In fact, the Large Magellanic Cloud (along with a second galaxy called the Small Magellanic Cloud) may actually be a satellite of the Milky Way Galaxy, the way that the moon is a satellite of the earth.

And unlike the majority of stars in our own galaxy, the Large Magellanic Cloud lies in a direction from earth that contains a minimum amount of interstellar particles through which astronomers must look. For example, there aren't many clouds to get in the way of our observations. Hence, the viewing is clearer in that direction than in most directions within our own galaxy.

Finally, the fact that the supernova lies within the LMC greatly simplifies the job of figuring out how far away it is. And that, in turn, simplifies the job of studying it.

As we saw earlier, the apparent brightness of a star in the sky depends on how far away it is. A star that appears very bright as viewed from earth may in fact be a very bright star, or it may just be unusually close to us. On the other hand, a dim star may be an unusually bright star that happens to be quite far away. It is sometimes difficult to tell which is which.

To simplify this confusing situation, astronomers use two different terms in discussing the brightness of stars: *apparent magnitude* and *absolute magnitude*. The magnitude of a star is its brightness, measured in numbers; the smaller the number, the brighter the star. The apparent magnitude is how bright the star looks to us from earth. The absolute magnitude is how bright it actually is—or, more precisely, how bright it would appear if viewed from a distance of 10 parsecs. (A *parsec* is a measure of distance equal to about 3.25 light-years, or 30 trillion kilometers.)

For example, the brightest star in the sky, Sirius, is so bright that its apparent magnitude is a negative number: −1.46. (Remember: the lower the magnitude number, the brighter the star.) As it happens, Sirius really is a bright star, so its absolute magnitude is also a negative number: −1.42. On the other hand, the star Alpha Centauri appears quite bright (apparent magnitude: 0.07) but is actually rather dim (absolute magnitude: 4.37). It appears bright only because it is unusually close to earth.

The brightest object in the sky is, obviously, the sun. Its apparent magnitude is a whopping −26.72. However, the sun's absolute magnitude is roughly the same as Alpha Centauri's: 4.86.

To figure the absolute magnitude of a star, we must know both its apparent magnitude and its distance from the earth. It's not hard to measure the apparent magnitude of a star or other object in the sky; there are instruments that will do this for us. Measuring distance, however, is a lot trickier. We can't simply shoot a measuring tape into the sky and count off the light-years. The stars are much too far away.

Over the years, astronomers have developed many clever methods of measuring the distance to stars and galaxies, but while these methods have allowed us to study the basic structure of the universe, they are most reliable with stars that are very close to us or with galaxies that are very far away. For individual stars at dis-

*the star Alpha Centauri
(the bright object in the center left)*

tances measurable in the dozens of light-years, our measurements are not always terribly accurate.

It is important that we know the distance to a supernova because then we can calculate its absolute magnitude, not only at the brightest point of the explosion but before the explosion (using old photographs of the star that exploded) and after the explosion. If a star in our galaxy exploded, astronomers might have trouble figuring out how far away it is. But the distance to the Large Magellanic Cloud was measured many years ago, and so the distance to the supernova was known immediately. Of course, the 170,000 light-year distance to the LMC is merely an average distance to any specific star within the cloud, but at that distance a few light-years either way makes little difference in how bright the star appears from earth.

How bright was Supernova 1987A? On the night that Shelton first photographed it, the apparent magnitude was between 5 and 4, not terribly bright by ordinary standards of visibility but very impressive when you consider that no ordinary star is visible at all over such a distance. (The Large Magellanic Cloud itself is visible to the naked eye to observers in the Southern Hemisphere, but it appears as a patch of nebulous fuzz

Before (top) and after photographs of Supernova 1987A. The before photo was taken in 1969; the after photo on February 26, 1987.

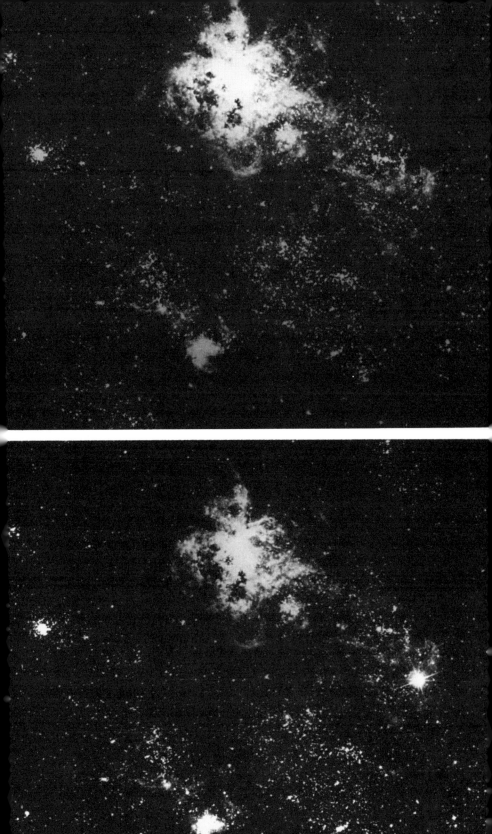

in the sky, with no individual star distinguishable from the others in the galaxy.) This corresponds to an absolute apparent magnitude of −18.4.

The initial assumption among astronomers was that this represented just the beginning of a brightening process that would eventually bring the supernova to an apparent magnitude of around 1, making it one of the brightest stars in the sky. However, this didn't happen. The star leveled off at about magnitude 4.5, then stayed there for several weeks. Eventually, however, it began brightening again, reaching a peak apparent magnitude of 2.88 on May 22, 1987.

The first thing astronomers wanted to know about 1987A, however, was not how bright it was but which star had exploded. Fortunately, the Large Magellanic Cloud had been photographed and mapped by astronomers long before the explosion. Finding out what star had exploded was a simple matter of examining pictures of the galaxy—or so it seemed at first. However, problems immediately arose.

Within hours, a star had been found within maps of the Large Magellanic Cloud in the same position where astronomers were now observing the supernova. It was called Sanduleak −69° 202. The numbers indicate its position in the sky and the name *Sanduleak* refers to the astronomer who had catalogued the star some years before.

Sanduleak was a giant star, about twenty times as massive as our sun. But it was a *blue giant* rather than a red giant, which meant that it was relatively small for a giant and not yet in the stage of its life at which conventional theory says that stars should explode. As we saw in Chapter Four, only stars many times the size of our sun end their lives in supernova explosions. But astronomers wondered why a star so small, relatively speaking, would have exploded at all.

It was difficult to verify the star's identity, however. The light from the supernova was so bright that any other stars in its vicinity were effectively lost in the glare; there was no way to tell if Sanduleak was still there or not. Then, as certain kinds of light from the supernova began to fade, a satellite-based telescope in orbit around the earth took pictures of the supernova that seemed to detect the light from Sanduleak still shining a little to one side of it. Astronomer Robert Kirschner, of the Harvard Smithsonian Center for Astrophysics, announced, on the basis of this observation, that Sanduleak was still in existence.

This stymied the astronomical community. Sanduleak was the only star in the proper position to be the progenitor of the supernova—or, rather, it was the only such star visible from earth. And while it was small as giant stars go, it was at least a giant. If the star that had exploded was too small to be seen from earth, then it would be far smaller than Sanduleak—and much too small to have become a supernova, according to all of the standard theories on the subject.

Could the theories be completely wrong? If astronomers had been appalled at the idea that a blue giant like Sanduleak may have been the progenitor of the supernova, they were even more appalled at the suggestion that it might *not* be! Fortunately, just as some astronomers had begun to retool their theories of stellar explosions, Kirschner reexamined the data from the satellite and discovered that the light being observed in the vicinity of the explosion was in the wrong position to be Sanduleak, and that no light from Sanduleak was visible after all. The big star was gone. Sanduleak was apparently the exploding star, just as astronomers had initially assumed.

Several weeks after the initial explosion astronomers were struck by yet another surprise. A strange companion object had appeared about 2 light-weeks—the distance that light travels in two weeks—from the supernova. Through the telescope, or more specifically, through a device known as a speckle interferometer, which can detect details too small to be seen with a normal telescope, the companion appears as a point of light. But it is a point of light that was not visible before the supernova explosion. And it is a very bright point of light, almost one-tenth as bright as the dazzling supernova itself, much too bright to be an ordinary star.

Could a second star have exploded in the LMC? Not likely; such a double supernova would be without precedent in the history of astronomy, a coincidence too fantastic to be seriously contemplated. Could the first supernova have triggered a second supernova? Once again, not likely. Astronomers know of no mechanism by which one supernova can cause another.

Perhaps the second object is merely a cloud of gas and dust reflecting the light of the supernova. This possibility has been seriously suggested, but such a cloud should have been blown apart by the shock wave from the explosion not long after it was first observed.

Another possible explanation for the strange companion offered thus far is this: The supernova explosion blew a large amount of solid matter into space, which is expanding outward from the site of the original star in a giant shell. (In fact, this shell will one day form a giant nebula in space much like the Crab Nebula.) At the middle of this expanding shell is a newly formed neutron star—hot, spinning rapidly, and shooting large amounts of hot particles into space in a form known as *plasma*. This plasma "jet" would be trapped by the expanding shell of matter, but a small amount of it might be able to leak through.

If the exploding star had a companion star in orbit around it—and most stars do—the companion star might have punched a hole in the shell as the shell expanded past it. The plasma jet might then be able to leak through this hole, like a thin stream of water shooting through a tiny hole in a rubber hose. The point of light that looks like a companion object through the telescope could actually be the leading edge of this hot plasma jet, rocketing outward into space. Computer calculations show that the jet would be traveling at the right speed to be in the observed position of the companion. However, only future calculations will show if this theory is correct.

6
BRIGHT LIGHTS, DARK MATTER

And what of the neutrinos from the supernova? Was the explosion of Sanduleak −69° 202 close enough to earth for astronomers to observe the outburst of neutrinos that it produced?

Yes, it was. But the results were oddly ambiguous. Astronomers learned a great deal, but they aren't yet sure what it was they learned.

In fact, the first neutrinos from the supernova arrived on earth the day before Ian Shelton observed its light. Around the world, neutrino detectors recorded tiny bursts of radiation that indicated several neutrino "events" had occurred within their huge vats of fluid. In Kamioka, Japan, the Kamiokande II neutrino detector recorded eleven neutrino events. A solar neutrino detector in Cleveland recorded eight events. In the French Alps, another detector may have registered five. These results tell us that the theory is correct at least to this extent: There really is a burst of neutrinos prior to a supernova.

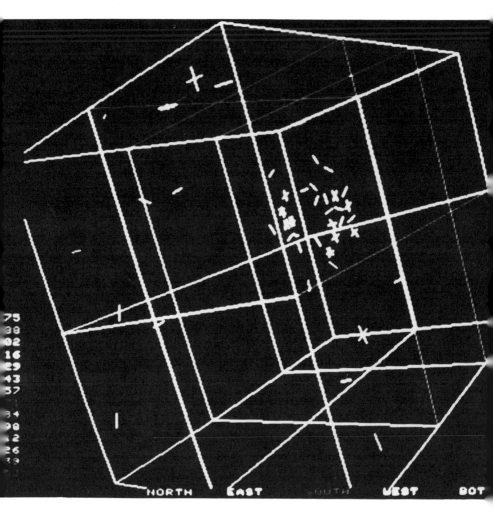

A computer display of a neutrino detected from Supernova 1987A. The display shows the outline of the detector, a tank of ultrapure water in a salt mine in Ohio. The broken "circle" in the center is due to a positron, which was produced with a neutron when an antineutrino from the supernova collided with a proton in the water.

It might seem surprising that an event 170,000 light-years away could produce enough neutrinos to be detected here on earth, considering that only a very small fraction of the neutrinos will interact with the detector at all. But the number of neutrinos released by the explosion is astonishing. Written down, it would be a one followed by fifty-eight zeroes! A human being 1 billion miles from the explosion would have been killed by the neutrino blast alone!

As you'll recall from Chapter Four, the precise timing of the neutrino events recorded by the detectors is important, because it can tell us whether or not the neutrino has any mass, that is, whether it weighs anything. If neutrinos move at the speed of light, then they are weightless; if they move more slowly than light, they must have at least a little weight.

The neutrinos did indeed show up before the light from the explosion, but this doesn't mean that the neutrinos have no weight at all. It just means that they are extremely light and *may* have no weight. Alas, we already know that, so the precise timing of the events—exactly how long they arrived before the light from the supernova—becomes of paramount importance.

Unfortunately, not all of the neutrino detectors agreed on this point. The Kamioka detector recorded the supernova events at about twenty-two hours before Shelton first observed the supernova. However, the neutrino detector in the French Alps registered the events four-and-a-half hours earlier than that!

There's no way that the neutrinos could have taken four-and-a-half hours to travel from the French Alps to Japan, so one of the detectors must have recorded a false series of events or gotten the time wrong for the real series. Previous problems with the detector in the Alps have led scientists to suspect it as the one producing the false results. The Japanese detector, on the other hand, is larger and therefore more accurate and has proven more reliable in the past.

The timing of the neutrino events tells us that neutrinos may indeed have no mass at all, but it doesn't prove this. Instead, it places an upward limit on the mass of the neutrino: about sixteen electron volts. (The electron volt, while actually a measure of energy, is used as a measure of extremely small amounts of mass.) The neutrino cannot have a greater mass than this, according to the observations of Supernova 1987A.

Sixteen electron volts is a very small mass, but as we saw in Chapter Four there are a huge number of neutrinos abroad in the universe, and even such a tiny amount of mass can add up. Why should this matter to scientists?

About 15 billion years ago, the universe was created in an immense explosion, an explosion that would make even the eruption of a supernova seem like the insignificant pop of a firecracker. We call this explosion the *Big Bang*. In this explosion all of the matter in the universe was flung outward from some unknown central point, eventually to coalesce into galaxies and stars. This early universe would have been filled with large, hot stars of the type that race through their life cycles in a few million years and explode as supernovas, so there would have been a lot more supernova explosions in the universe then than there are today.

The universe probably looks a lot different today, but it is still expanding. Astronomers can actually see this expansion when examining faraway galaxies, all of which are rushing away from our local group of galaxies and from each other.

Will this expansion ever end? Will the universe keep on expanding forever? Or will the expansion slow down and eventually stop altogether?

And, if it stops altogether, will the gravity of all the stars and galaxies in the universe eventually cause it to collapse again, perhaps triggering another Big Bang and yet another expanding universe?

No one knows. Whether the expansion continues or stops depends on how much gravity is produced by all of the matter in the universe, that is, whether there is enough gravity to put a brake on the expansion. If there is enough matter to halt the expansion we say that the universe is *closed* and will eventually stop expanding. If there isn't enough matter to halt the expansion we say that the universe is *open* and that it will go on expanding forever. But determining how much gravity is produced by all the matter in the universe means that we must first determine how much matter there is in the universe, and that is easier said than done.

When astronomers add up the mass of all the objects that they see in the heavens they come up with a figure that is roughly one-tenth to one-hundredth the amount of mass required to close the universe, that is, to eventually halt the expansion. However, there is evidence that there is more matter in the universe than astronomers can see through their telescopes. For instance, the manner in which galaxies rotate indicates that they are surrounded by large invisible "haloes" and that these haloes may contain from ten to one hundred times as much matter as do their visible portions. In fact, these haloes may contain enough matter to close the universe.

The matter in these haloes is referred to by astronomers as *dark matter* or the *missing mass*. Obviously, it is not ordinary matter. It does not emit light or any kind of detectable radiation, like the hot matter in stars. On the other hand, it does not absorb the light from the galaxy that it surrounds, the way ordinary cool matter would. It is effectively invisible and seems to interact with ordinary matter only through its gravity.

Neutrinos, you will recall, are extremely reluctant to interact with ordinary matter. So neutrinos would seem to be a reasonable candidate for the dark matter in these galactic haloes. However, if neutrinos have no mass, as Wolfgang Pauli suggested, then they will produce no gravity, and the dark matter surrounding galaxies is clearly producing gravity. It is only through its gravitational interaction with the rest of the galaxy that we know it exists. On the other hand, if neutrinos have only a small amount of mass, then with their large numbers they could be producing the gravity of the galactic haloes.

And that would mean that there is from ten to one hundred times more gravity-producing matter in the universe than astronomers have observed. If neutrinos have mass, then the universe might well be closed.

This is why astronomers are so carefully monitoring the results of the neutrino observations of Supernova 1987A. The precise timing of those observations relative to the visible light observations of the supernova could have large implications in terms of the future existence of the universe.

If the neutrinos arrived here as quickly as the light, then the universe is quite possibly open and will go on expanding forever. One day billions of years in the future all of the stars will run out of energy and there will not be enough hydrogen to create new stars through gravitational contraction. The theories of physics tell us that even the particles that make up atoms will eventually fall apart with age, and all of the matter in the universe will be reduced to its smallest constituent parts. The stars will vanish, but the universe will expand forever.

If the neutrinos arrived here a little more slowly than the light, then the universe is quite possibly closed and will eventually collapse. If so, there may ultimately be another Big Bang and another universe, which in turn will collapse to produce yet another Big Bang, ad infinitum.

Five billion years ago the formation of our solar system may have been triggered by the explosion of a supernova. Most of the atoms in your body were formed in the hearts of supernovas. And now the fate of the universe—or, at least, our knowledge of that fate—may hinge on a few seemingly trivial observations of a supernova.

Is it any wonder that, for a few months in 1987, the attention of astronomers all over the world was focused on a single star that exploded a long time ago in a galaxy not so very far away? Supernova 1987A is, for some astronomers, the event of a lifetime. It will continue providing information for years, as astronomers observe the expanding shell of matter that surrounds it. Eventually, if we are lucky, we may detect signals from the neutron star that may have formed at the heart of the explosion. Given time, scientists may be able to adapt their theories of supernovas to accommodate the fact that Sanduleak was a blue giant rather than a red giant. Could the evolution of stars follow a different path in small irregular galaxies such as the Large Magellanic Cloud than in larger spiral galaxies such as our own?

Until the next time a nearby star explodes—some astronomers, for instance, predict that Betelgeuse may explode sometime in the next million years or so and some more distant star may explode much sooner—Supernova 1987A, first photographed from a remote peak in Chile on a windy February night, will provide us with the best information we have ever had about a phenomenon that astronomers and other stargazers have been wondering about for as long as there has been a human race.

GLOSSARY

ABSOLUTE MAGNITUDE—A number representing the magnitude (brightness) of a star as it would appear at a distance of 10 parsecs from earth.

APPARENT MAGNITUDE—A number representing the magnitude (brightness) of a star as it appears when viewed from earth.

ASTROLOGY—A pseudoscience developed in ancient times by astronomers that attempted to explain how events on earth were influenced by heavenly bodies such as the planets.

BIG BANG—The explosion that astronomers believe represents the creation of the universe.

BINARY STARS—A pair of stars in orbit around one another.

BLACK DWARF—A collapsed star; its atoms are held apart by electron degeneracy pressure; the star is black because most of the heat (and thus light) of its former fusion reactions has dissipated into space.

BLACK HOLE—The remains of an extremely large star in which gravitational collapse continues indefinitely, causing the stellar remains to approach infinite smallness and infinite density. Black holes produce so much intense gravitational attraction that not even light is able to escape from its sphere of influence.

BLUE GIANT—A giant star that is hotter and younger than a red giant. Astronomers believe that Sanduleak−69° 202, the star that exploded in the Large Magellanic Cloud in 1987, was a blue giant.

CHAIN REACTION—The process by which nuclear fission sustains itself; the protons ejected by one fissioning nucleus promote the fissioning of additional nuclei. Atomic bombs use fission to produce their energy.

CLOSED UNIVERSE—Theoretical model of the universe that says the expanding universe will eventually reverse itself, due to gravitational attraction, and collapse.

CRITICAL MASS—The amount of a fissionable substance that must be brought together in order for a self-sustaining chain reaction to take place.

DARK MATTER—"Invisible" matter that astronomers believe may constitute at least 90 percent of the matter in the universe; possibly concentrated in the outer regions of galaxies, the so-called *galactic haloes*.

DEGENERATE MATTER—Matter in which the normal atomic structures have been crushed by intense gravitational forces.

ECLIPSING BINARY—A binary star system in which one star occasionally moves in front of the other star, as seen by an observer on earth.

ELECTRON—One of the extremely small particles of which atoms are made. Electrons have a negative charge.

ELECTRON DEGENERACY PRESSURE—The force created by the "refusal" of electrons to approach one another more closely than they would ordinarily within the electron shell of an atom.

ETHER—Substance from which Aristotle believed stars and all other heavenly objects were made; root for the modern word *ethereal.*

FISSION—The process by which large unstable atoms (or, more specifically, atomic nuclei) split apart to form smaller atoms (or, rather, nuclei), releasing large amounts of energy as they do so.

FUSION—The process by which atoms (or, rather, nuclei) come together ("fuse") to form larger atoms (nuclei), releasing large amounts of energy as they do so. This is the process that powers the sun.

HYDROGEN FUSION—The fusion of hydrogen nuclei to form helium nuclei.

INTERSTELLAR MEDIUM—The thin "haze" of gas and dust particles that fills the otherwise empty space between the stars.

MAGNITUDE—Measure of the brightness of a star in magnitude numbers—the smaller the number, the brighter the star.

MISSING MASS—Matter that astronomers believe may exist, but that they cannot see; required to "close the universe," that is, to assure that the expanding universe will eventually reverse itself; may be the same as the "dark matter."

MULTIPLE-STAR SYSTEM—System in which two or more stars are orbiting about one another.

NEUTRINOS—Extremely tiny particles created during

certain nuclear reactions, such as fusion; believed to have either no mass or very little mass.

NEUTRON—One of the tiny particles of which atoms are made. Neutrons have no electrical charge and reside in the atom's nucleus.

NEUTRON STAR—An extremely dense collapsed star in which the electrons, protons, and neutrons have been converted almost entirely into neutrons.

NOVA—A star that flares to sudden intense brightness; from the Latin *stella nova* ("new star"), because novas were once incorrectly believed to be stars that appeared where none had existed previously.

NUCLEAR FORCE—The force that holds neutrons and protons together within an atomic nucleus, and that holds neutrons apart within a neutron star.

NUCLEUS—The central portion of the atom, in which are found the neutrons and protons.

OPEN UNIVERSE—Theoretical model of the universe that says the universe will continue expanding forever.

PARSEC—A measure of astronomical distance equal to about 3.25 light-years, or 30 trillion kilometers.

PLASMA—An extremely hot, energetic, gaseous form of matter.

POPULATION I STARS—Stars that are rich in heavy elements formed in earlier supernova explosions.

POPULATION II STARS—Stars that are poor in heavy elements.

PROPER MOTION—The apparent motion of a star in the sky relative to other stars.

PROTON—One of the small particles of which atoms are made. Protons have a positive electrical charge and reside in the nucleus.

PULSAR—Short for "pulsating star"; a rotating neutron star that sends rhythmic electromagnetic signals into space where they can be detected by radio astronomers on earth.

RED GIANT—Star in the last stages of its evolution, swollen to great size by the heat of helium fusion at its core.

RETROGRADE MOTION—The apparent reversal of a planet's motion against the background stars as viewed by an observer on earth, caused by the earth's own motion relative to the other planet.

SUPERNOVA—A particularly bright nova, caused either by the explosion of a very large old star or the sudden vaporization of materials drawn off of a star onto the surface of its binary companion, usually a white dwarf.

TYPE I SUPERNOVA—A supernova explosion resulting from the drawing off of material from a star onto its binary companion, usually a white dwarf.

TYPE II SUPERNOVA—A supernova explosion resulting from the death of an old, very large star.

VARIABLE STARS—Stars that change apparent magnitude in a cyclical manner, becoming brighter, then darker, then brighter again.

WHITE DWARF—A collapsed star, its atoms held apart by electron degeneracy pressure, which still glows from the residual heat generated by fusion and the gravitational energy of collapse.

SOURCES USED

Books

Asimov, Isaac. *The Exploding Suns.* E. P. Dutton, New York: 1985.

Clark, David H. *Superstars: How Stellar Explosions Shape the Destiny of Our Universe.* McGraw-Hill, New York: 1984.

Cooke, Donald A. *The Life and Death of Stars.* Crown, New York: 1985.

Friedlander, Michael W. *Astronomy: From Stonehenge to Quasars.* Prentice Hall, New Jersey: 1985.

Greenstein, George. *Frozen Star.* Freundlich Books, New York: 1983.

Illingworth, Valerie, ed. *The Facts on File Dictionary of Astronomy.* Facts on File, New York: 1985.

Articles

Bahcall, J. N., A. Dar, & T. Piran. "Neutrinos from the Recent LMC Supernova," *Nature,* March 12, 1987, Vol. 326, p. 135.

Campbell, Philip. "Supernova Almost on the Doorstep," *Nature*, March 5, 1987, Vol. 326, p. 11.

_____. "Which Star Went Bang?" *Nature*, April 9, 1987, Vol. 326, p. 543.

"Damp Squib Supernova Excites Particle Physicists," *New Scientist*, March 12, 1987, p. 24.

Garwin, Laura. "Conflicting Signals from LMC Supernova," *Nature*, March 12, 1987, Vol. 326, p. 121.

Henbest, Nigel. "Exploding Star Startles Astronomers," *New Scientist*, March 5, 1987, p. 14.

_____. "New Observations Support Supernova Theories," *New Scientist*, April 7, 1987, p. 26.

Hillebrandt, W., P. Hoflich, J. W. Truran & A. Weiss. "Explosion of a Blue Supergiant: A Model for Supernova SN 1987A," *Nature*, June 18, 1987, Vol. 327, p. 597.

Lemonick, Michael D. "Supernova!" *Time*, March 23, 1987, Vol. 129, No. 12, p. 60.

Lindley, David. "Neutrinos from the Supernova," *Nature*, March 19, 1987, Vol. 326, p. 239.

_____. "A Quiet Supernova?" *Nature*, March 26, 1987, Vol. 326, p. 328.

_____. "Supernova Examined by Computer Model," *Nature*, April 16, 1987, Vol. 326, p. 638.

Morgan, Charles. "Australian View of the Supernova," *Nature*, April 2, 1987, Vol. 326, p. 440.

Sawyer, Kathy. "Super Lenses in Chile Gaze at Supernova," *The Washington Post*, March 15, 1987, Vol. 110, No. 100, p. A–1.

_____. "Star Explosion in Nearby Galaxy Revolutionizing Astronomy," *The Washington Post*, March 22, 1987, Vol. 110, No. 107, p. A–3.

_____. "Cosmic Fireworks May Be in Store from Supernova," *The Washington Post*, June 22, 1987, Vol. 110, No. 199, p. A–3.

"Supernova Sets Limits on Neutrino Mass," *New Scientist,* March 19, 1987, p. 24.

Talcott, Richard. "A Burst of Discovery: The First Days of Supernova 1987A," *Astronomy,* June 1987, Vol. 15, No. 6, p. 90.

Tierney, John. "Exploding Star Contains Atoms of Elvis Presley's Brain," *Discover,* July 1987, Vol. 8, No. 8, p. 46.

Waldrop, M. Mitchell. "Sighting of a Supernova," *Science,* March 26, 1987, Vol. 235, p. 1143.

———. "The Supernova 1987A Shows a Mind of Its Own—and a Burst of Neutrinos," *Science,* March 13, 1987, Vol. 235, p. 1322.

———. "Supernova Neutrinos at IMB," *Science,* March 20, 1987, Vol. 235, p. 1461.

———. "Supernova 1987A: Notes from All Over," *Science,* May 2, 1987, Vol. 236, p. 522.

———. "Supernova 1987A: A Mysterious Stranger," *Science,* July 3, 1987, Vol. 237, p. 25.

RECOMMENDED READING

Asimov, Isaac. *The Exploding Suns* (Dutton, New York: 1985). An extremely clear explanation and history of supernovas by the dean of science writers. Written before Supernova 1987A, however.

―――――. *The Universe* (Walker, New York: 1979). A well-written introduction to the science of astronomy.

Clark, David H. *Superstars* (McGraw-Hill, New York: 1984). The story of supernovas by an astronomer who has studied them throughout much of his career.

Cooke, Donald A. *The Life and Death of Stars* (Crown, New York: 1985). Everything you ever wanted to know about the life and death of stars, vividly illustrated and lucidly written.

Lampton, Christopher. *Black Holes and Other Secrets of the Universe* (Franklin Watts, New York: 1980). An introduction to black holes, neutron stars, supernovas, and quasars for young readers, by the author of this book.

INDEX

Absolute magnitude, 97–98
Alpha Centauri, 98
Apparent magnitude, 97–98
Aristotle, 28, 31, 35, 36, 49
Astrology, 27–28
Astronomical Telegrams, Central Bureau for, 16
Astronomy, 28, 31, 35–36
Atom, 56–59, 61

Baade, Walter, 83–84
Bell, Jocelyn, 76
Betelgeuse, 68, 113–114
Bethe, Hans, 61

Big Bang theory, 110–113
Binary stars, 27
Black dwarf, 70
Black hole, 78, 80
Blue giant, 102
Brahe, Tycho. *See* Tycho Brahe

Chain reaction, 59
Chandrasekhar, Subrahmanyan, 76, 84
Christian Church, 31, 40
Closed universe, 111
Comets, 35–36
Copernicus, Nicolaus, 40, 45, 49
Crab Nebula, 78, 85
Critical mass, 59

Dark matter, 111–113
Darwin, Charles, 51
Degenerate matter, 70
De Stella Nova (Tycho), 40, 45
Duhalde, Oscar, 9, 13–16

Earth, 22–23
Eclipsing binary, 27
Eddington, Sir Arthur, 76, 84
Egyptians, 37
Einstein, Albert, 54, 56
Electron, 57
Electron degeneracy pressure, 70
Energy, equation, 56
Ether, 49

Fission, 58–59
Fusion, 61, 63

Galaxies, 83, 111–113
Geologists, 51
Gravity, 53, 111
Greeks, 28, 31
Gum, Colin S., 85
Gum Nebula, 85

Halley, Edmund, 36
Halley's Comet, 35
Haloes, galactic, 111–113
Helium, 67

Helmholtz, Ludwig von, 51, 53–54, 67
Hewish, Antony, 76
Hydrogen, 67
Hydrogen fusion, 61, 63, 67

International Ultraviolet Explorer (satellite), 17
Interstellar medium, 65, 67

Kamiokande II neutrino detector, 107, 109–110
Kepler, Johannes, 45
Kirschner, Robert, 103

Large Magellanic Cloud (LMC), 14, 16–17, 19, 95–114
Las Campanas Observatory, 9, 95
Light-year, 98, 100
LMC. *See* Large Magellanic Cloud

Magnitude, of star, 97–98
Michanowsky, George, 37–38
Middle Ages, 31
Milky Way Galaxy, 9, 16, 83–84, 95, 97
Missing mass, 111–113
Multiple-star system, 27

Natural selection, 51
Neutrinos, 87, 89–90,
  92–93
  observed in Supernova 1987A, 107,
  109–114
Nuetrino telescope, 89–90, 92
Neutron, 57
Neutron star, 75–78
Night sky, 22, 27
Nova, 36–37
  compared with supernova, 85–86
Nuclear force, 75
Nucleus, 57–58

Open universe, 111
Orbit, of earth, 22

Paleontologists, 51
Parsec, 98
Pauli, Wolfgang, 87, 89, 92, 112
Plasma, 105
Plutonium atom, 58–59
Population I and II stars, 75
Proper motion, 23, 27
Proton, 57
Pulsar, 76–78

Red giant, 68, 70
Retrograde motion, 23

Rotation, of earth, 22

Sanduleak, 102–104
Shelton, Ian, 14, 16, 19, 95
Sirius, 98
Sky, 21–23, 27–28, 31, 35–36
Small Magellanic Cloud, 97
Solar Max (satellite), 16
Space, outer, 65
"Special Theory of Relativity" (Einstein), 54, 56
Speckle interferometer, 104
Stars, 21–22
  black dwarfs, 70
  black hole, 78, 80
  formation of, 67
  magnitude of, 97–98
  measuring distance of, 98, 100
  neutron star, 75–78
  Population I and II stars, 75
  pulsar, 76–78
  red giants, 68, 70
  variable stars, 27
  white dwarfs, 70
Sumerians, 37
Sun, 22–23, 53–54, 56, 67–68

Supernova, 37–38, 40, 45, 47, 63, 74
  and neutrinos, 89–90, 92–93
  theories on, 84
  types of, 81, 83, 85–87
  See also Supernova 1987A
Supernova 1987A, 16–17, 19, 97
  magnitude of, 100, 102
  neutrino events in, 107, 109–114
  and Sanduleak, 102–104

Tarantula Nebula, 13
Telescope, 45
  neutrino, 89–90, 92
Tycho Brahe, 40, 45, 74
Type I and II supernovas, 86–87

Universe:
  Big Bang Theory of, 110–113
  views on, 28, 31, 35
Uranium atom, 58–59

Variable stars, 27
Vela supernova, 37
Voyager 2 (space probe), 17

White dwarf, 70, 86, 87

Zwicky, Fritz, 83–84